Student Solutions

John Garlow

BEGINNING ALGEBRA

Third Edition

John Tobey

Jeffrey Slater

North Shore Community College
Beverly, Massachusetts

Prentice Hall, Englewood Cliffs, NJ 07632

Production Editor: *Rosemary Madigan*
Acquisitions Editor: *Melissa Acuña*
Supplement Acquisitions Editor: *Audra Walsh*
Production Coordinator: *Alan Fischer*

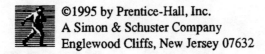

©1995 by Prentice-Hall, Inc.
A Simon & Schuster Company
Englewood Cliffs, New Jersey 07632

Printed in the United States of America

10 9 8 7 6 5 4 3 2 1

ISBN 0-13-310053-7

Prentice-Hall International (UK) Limited, *London*
Prentice-Hall of Australia Pty. Limited, *Sydney*
Prentice-Hall Canada Inc., *Toronto*
Prentice-Hall Hispanoamericana, S.A., *Mexico*
Prentice-Hall of India Private Limited, *New Delhi*
Prentice-Hall of Japan, Inc., *Tokyo*
Simon & Schuster Asia Pte. Ltd., *Singapore*
Editora Prentice-Hall do Brasil, Ltda., *Rio de Janeiro*

TABLE OF CONTENTS

Pretest Chapter 0

1. $\dfrac{21}{27} = \dfrac{3 \times 7}{3 \times 9} = \dfrac{7}{9}$

3. $\dfrac{15}{4} = 4\overline{)15} = 3\dfrac{3}{4}$
$\quad \dfrac{12}{3}$

5. $\dfrac{3}{7} = \dfrac{?}{14} \Rightarrow \dfrac{3 \times 2}{7 \times 2} = \dfrac{6}{14}$

7. $\dfrac{3}{8}, \dfrac{5}{6}, \dfrac{7}{15}:$ $\quad 8 = 2 \cdot 2 \cdot 2$
$\qquad\qquad\qquad 6 = \quad\quad 2 \cdot 3$
$\qquad\qquad\qquad 15 = \qquad\quad 3 \cdot 5$
$\qquad\quad \text{LCD} = 2 \cdot 2 \cdot 2 \cdot 3 \cdot 5 = 120$

9. $\dfrac{5}{48} + \dfrac{5}{16} = \dfrac{5}{48} + \dfrac{5 \cdot 3}{16 \cdot 3} = \dfrac{5}{48} + \dfrac{15}{48} = \dfrac{20}{48} = \dfrac{4 \cdot 5}{4 \cdot 12} = \dfrac{5}{12}$

11. $3\dfrac{1}{5} - 1\dfrac{3}{8} = \dfrac{16}{5} - \dfrac{11}{8} = \dfrac{128}{40} - \dfrac{55}{40} = \dfrac{73}{40} = 1\dfrac{33}{40}$

13. $2\dfrac{4}{5} \times 3\dfrac{3}{4} = \dfrac{14}{5} \times \dfrac{15}{4} = \dfrac{2 \cdot 7}{5} \times \dfrac{3 \cdot 5}{2 \cdot 2} = \dfrac{21}{2} = 10\dfrac{1}{2}$

15. $2\dfrac{1}{3} \div 3\dfrac{1}{4} = \dfrac{7}{3} \div \dfrac{13}{4} = \dfrac{7}{3} \times \dfrac{4}{13} = \dfrac{28}{39}$

17. $\dfrac{5}{9} = 9\overline{)5.0} = 0.\overline{5}$
$\qquad \dfrac{45}{50}$
$\qquad \dfrac{45}{\ }$

19. $\quad 3.28$
$\quad \underline{\times\ 0.63}$
$\qquad 984$
$\quad \underline{1968\ \ }$
$\quad 2.0664$

21. $\quad 12.130$
$\quad \underline{-\ \ 9.884}$
$\qquad 2.246$

23. $0.006 = 0.6\%$

25. $2\% = 0.02$

27. What percent of 500 is 36?
$\qquad \dfrac{36}{500} = 0.072 = 7.2\%$

29. $7386 \times 2856 \approx 7000 \times 3000 = 21,000,000$

31. Total gallons $= 3.7 + 10 + 12 + 4.9 = 30.6$
$\qquad \dfrac{870}{30.6} \approx 28.4$ miles per gallon

Exercises 0.1

1. $\dfrac{12}{13}:$ numerator $= 12$

3. When two or more numbers are multiplied, each number that is multiplied is called a factor. In 2×3, 2 and 3 are factors.

5. $2\dfrac{3}{5}$

7. $\dfrac{35}{49} = \dfrac{7 \times 5}{7 \times 7} = \dfrac{5}{7}$

9. $\dfrac{12}{60} = \dfrac{12 \times 1}{12 \times 5} = \dfrac{1}{5}$

11. $\dfrac{75}{15} = \dfrac{15 \times 5}{15 \times 1} = 5$

13. $\dfrac{39}{7} = 7\overline{)39} = 5\dfrac{4}{7}$
$\qquad \dfrac{35}{4}$

15. $\dfrac{125}{4} = 4\overline{)125} = 31\dfrac{1}{4}$
$\qquad \dfrac{12}{\ }$
$\qquad \ \ \overline{5}$
$\qquad \ \ \dfrac{4}{1}$

17. $\dfrac{89}{6} = 6\overline{)89} = 14\dfrac{5}{6}$
$\dfrac{6}{}$
$\overline{29}$
$\underline{24}$
5

19. $2\dfrac{6}{7} = \dfrac{(2\times 7)+6}{7} = \dfrac{14+6}{7} = \dfrac{20}{7}$

21. $4\dfrac{3}{4} = \dfrac{(4\times 4)+3}{4} = \dfrac{16+3}{4} = \dfrac{19}{4}$

23. $13\dfrac{1}{6} = \dfrac{(13\times 6)+1}{6} = \dfrac{78+1}{6} = \dfrac{79}{6}$

25. $\dfrac{5}{7} = \dfrac{?}{21} \Rightarrow \dfrac{5\times 4}{7\times 4} = \dfrac{20}{28}$

27. $\dfrac{2}{7} = \dfrac{?}{21} \Rightarrow \dfrac{2\times 3}{7\times 3} = \dfrac{6}{21}$

29. $\dfrac{13}{17} = \dfrac{?}{51} \Rightarrow \dfrac{13\times 3}{17\times 3} = \dfrac{39}{51}$

31. $\dfrac{75}{105} = \dfrac{15\times 5}{15\times 7} = \dfrac{5}{7}$

33. $\dfrac{7}{8} = \dfrac{?}{136} \Rightarrow \dfrac{7\times 17}{8\times 17} = \dfrac{119}{136} \Rightarrow 119$ correct

35. $\dfrac{3}{1+3+1} = \dfrac{3}{5}$ is raisins

37. Dunsay: $\dfrac{15}{25} = \dfrac{5\times 3}{5\times 5} = \dfrac{3}{5}$

Banks: $\dfrac{12}{30} = \dfrac{6\times 2}{6\times 5} = \dfrac{2}{5}$

Re: $\dfrac{28}{35} = \dfrac{7\times 4}{7\times 5} = \dfrac{4}{5}$

Exercises 0.2

1. 8 is the LCD of $\dfrac{3}{4}$ and $\dfrac{5}{8}$ because 8 is exactly divisible by 4.

3. $\dfrac{1}{12}$ and $\dfrac{5}{9}$: $\quad 12 = 2\cdot 2\cdot 3$
$9 = 3\cdot 3$
$\text{LCD} = 2\cdot 2\cdot 3\cdot 3 = 36$

5. $\dfrac{7}{10}$ and $\dfrac{1}{4}$: $\quad 10 = 2\cdot 5$
$4 = 2\cdot 2$
$\text{LCD} = 2\cdot 2\cdot 5 = 20$

7. $\dfrac{5}{63}$ and $\dfrac{5}{21}$: $\quad 63 = 3\cdot 3\cdot 7$
$21 = 3\cdot 7$
$\text{LCD} = 3\cdot 3\cdot 7 = 63$

9. $\dfrac{1}{18}$ and $\dfrac{13}{30}$: $\quad 18 = 2\cdot 3\cdot 3$
$30 = 2\cdot 3\cdot 5$
$\text{LCD} = 2\cdot 3\cdot 3\cdot 5 = 90$

11. $\dfrac{3}{11} + \dfrac{2}{11} + \dfrac{4}{11} = \dfrac{3+2+4}{11} = \dfrac{9}{11}$

13. $\dfrac{8}{17} - \dfrac{5}{17} = \dfrac{8-5}{17} = \dfrac{3}{17}$

15. $\dfrac{5}{6} + \dfrac{8}{15} = \dfrac{5\cdot 5}{6\cdot 5} + \dfrac{8\cdot 2}{15\cdot 2} = \dfrac{25}{30} + \dfrac{16}{30} = \dfrac{41}{30}$

17. $\dfrac{5}{14} - \dfrac{1}{4} = \dfrac{5\cdot 2}{14\cdot 2} - \dfrac{1\cdot 7}{4\cdot 7} = \dfrac{10}{28} - \dfrac{7}{28} = \dfrac{3}{28}$

19. $\dfrac{2}{9} + \dfrac{5}{6} = \dfrac{2\cdot 2}{9\cdot 2} + \dfrac{5\cdot 3}{6\cdot 3} = \dfrac{4}{18} + \dfrac{15}{18} = \dfrac{19}{18} = 1\dfrac{1}{18}$

21. $\dfrac{4}{5} - \dfrac{3}{8} = \dfrac{4\cdot 8}{5\cdot 8} - \dfrac{3\cdot 5}{8\cdot 5} = \dfrac{32}{40} - \dfrac{15}{40} = \dfrac{17}{40}$

23. $\dfrac{9}{8} + \dfrac{7}{12} = \dfrac{9\cdot 3}{8\cdot 3} + \dfrac{7\cdot 2}{12\cdot 2} = \dfrac{27}{24} + \dfrac{14}{24} = \dfrac{41}{24} = 1\dfrac{17}{24}$

25. $\dfrac{7}{18} + \dfrac{1}{12} = \dfrac{7\cdot 2}{18\cdot 2} + \dfrac{1\cdot 3}{12\cdot 3} = \dfrac{14}{36} + \dfrac{3}{36} = \dfrac{17}{36}$

27. $\dfrac{2}{3} + \dfrac{7}{12} = \dfrac{2\cdot 4}{3\cdot 4} + \dfrac{7}{12} = \dfrac{8}{12} + \dfrac{7}{12} = \dfrac{15}{12} = \dfrac{5}{4} = 1\dfrac{1}{4}$

29. $\dfrac{1}{15} + \dfrac{7}{12} = \dfrac{1\cdot 4}{15\cdot 4} + \dfrac{7\cdot 5}{12\cdot 5} = \dfrac{4}{60} + \dfrac{35}{60} = \dfrac{39}{60} = \dfrac{13}{20}$

31. $\dfrac{23}{25} - \dfrac{13}{15} = \dfrac{23\cdot 3}{25\cdot 3} - \dfrac{13\cdot 5}{15\cdot 5} = \dfrac{69}{75} - \dfrac{65}{75} = \dfrac{4}{75}$

33. $5\dfrac{1}{6} + 3\dfrac{1}{4} = \dfrac{31}{6} + \dfrac{13}{4} = \dfrac{62}{12} + \dfrac{39}{12} = \dfrac{101}{12} = 8\dfrac{5}{12}$

35. $6\dfrac{2}{3} + \dfrac{3}{4} = \dfrac{20}{3} + \dfrac{3}{4} = \dfrac{80}{12} + \dfrac{9}{12} = \dfrac{89}{12} = 7\dfrac{5}{12}$

37. $6\dfrac{3}{8} - 2\dfrac{3}{4} = \dfrac{51}{8} - \dfrac{11}{4} = \dfrac{51}{8} - \dfrac{22}{8} = \dfrac{29}{8} = 3\dfrac{5}{8}$

39. $4\dfrac{7}{12} - 1\dfrac{5}{6} = \dfrac{55}{12} - \dfrac{11}{6} = \dfrac{55}{12} - \dfrac{22}{12} = \dfrac{33}{12} = 2\dfrac{9}{12} = 2\dfrac{3}{4}$

41. $3\dfrac{1}{7} + 4\dfrac{1}{3} = \dfrac{22}{7} + \dfrac{13}{3} = \dfrac{66}{21} + \dfrac{91}{21} = \dfrac{157}{21} = 7\dfrac{10}{21}$

43. $14\dfrac{1}{9} - 12\dfrac{5}{9} = \dfrac{127}{9} - \dfrac{113}{9} = \dfrac{14}{9} = 1\dfrac{5}{9}$

45. $3\dfrac{1}{8} + 2\dfrac{2}{3} + 4\dfrac{1}{2} = \dfrac{25}{8} + \dfrac{8}{3} + \dfrac{9}{2} = \dfrac{75}{24} + \dfrac{64}{24} + \dfrac{108}{24}$

$= \dfrac{247}{24} = 10\dfrac{7}{24}$ miles

47. $8\dfrac{1}{2} - 2\dfrac{2}{3} - 1\dfrac{3}{4} = \dfrac{17}{2} - \dfrac{8}{3} - \dfrac{7}{4} = \dfrac{102}{12} - \dfrac{32}{12} - \dfrac{21}{12}$

$= \dfrac{49}{12} = 4\dfrac{1}{12}$ hours

49. height $= 2 + \dfrac{1}{2} + 3\dfrac{1}{2} + \dfrac{1}{2} + 3\dfrac{1}{2} + \dfrac{1}{2} + 1\dfrac{1}{2}$

$= 12$ inches

Cumulative Review Problems

51. $\dfrac{36}{44} = \dfrac{9 \cdot 4}{11 \cdot 4} = \dfrac{9}{11}$

53. $\dfrac{136}{7} = 7\overline{)136} = 19\dfrac{3}{7}$
    $\;\;\dfrac{}{66}$
    $\dfrac{63}{3}$

Putting Your Skills to Work

1. (a) Plates $= \left[8' - \left(3" + 1\dfrac{1}{2}" \right) \right] - 7'4\dfrac{1}{2}"$

$= 7'7\dfrac{1}{2}" - 7'4\dfrac{1}{2}" = 3"$

Each plate is $1\dfrac{1}{2}"$

(b) Width $= 16" + 16" = 32"$

(c) Width $= 7'8\dfrac{1}{2}" - 2\left(12\dfrac{1}{2}" \right) - 2(16") - 2\left(1\dfrac{3}{4}" \right)$

$= 92\dfrac{1}{2}" - 60\dfrac{1}{2}"$

$= 32"$

Exercises 0.3

1. $\dfrac{3}{5} \times \dfrac{2}{11} = \dfrac{3 \cdot 2}{5 \cdot 11} = \dfrac{6}{55}$

3. $\dfrac{25}{3} \times \dfrac{6}{5} = \dfrac{5 \cdot 5}{3} \times \dfrac{2 \cdot 3}{5} = 5 \cdot 2 = 10$

5. $8 \times \dfrac{3}{7} = \dfrac{8}{1} \times \dfrac{3}{7} = \dfrac{24}{7} = 3\dfrac{3}{7}$

7. $1\dfrac{1}{3} \times 2\dfrac{1}{2} = \dfrac{4}{3} \times \dfrac{5}{2} = \dfrac{2 \cdot 2}{3} \times \dfrac{5}{2} = \dfrac{10}{3} = 3\dfrac{1}{3}$

9. $3\dfrac{1}{3} \times \dfrac{3}{4} \times 2 = \dfrac{10}{3} \times \dfrac{3}{4} \times \dfrac{2}{1} = \dfrac{2 \cdot 5}{3} \times \dfrac{3}{2 \cdot 2} \times \dfrac{2}{1} = 5$

11. $\dfrac{5}{3} \div \dfrac{5}{2} = \dfrac{5}{3} \times \dfrac{2}{5} = \dfrac{2}{3}$

13. $\dfrac{3}{7} \div 3 = \dfrac{3}{7} \div \dfrac{3}{1} = \dfrac{3}{7} \times \dfrac{1}{3} = \dfrac{1}{7}$

15. $10 \div \dfrac{5}{7} = \dfrac{10}{1} \times \dfrac{7}{5} = \dfrac{2 \cdot 5}{1} \times \dfrac{7}{5} = 14$

17. $\dfrac{\frac{5}{2}}{\frac{2}{3}} = \dfrac{5}{2} \times \dfrac{3}{2} = \dfrac{15}{4} = 3\dfrac{3}{4}$

19. $1\dfrac{3}{7} \div 6\dfrac{1}{4} = \dfrac{10}{7} \div \dfrac{25}{4} = \dfrac{10}{7} \times \dfrac{4}{25} = \dfrac{2 \cdot 5}{7} \times \dfrac{4}{5 \cdot 5} = \dfrac{8}{35}$

21. $\dfrac{2}{1\frac{1}{5}} = \dfrac{2}{\frac{6}{5}} = \dfrac{2}{1} \times \dfrac{5}{6} = \dfrac{2}{1} \times \dfrac{5}{2 \cdot 3} = \dfrac{5}{3} = 1\dfrac{2}{3}$

23. $\dfrac{6}{5} \times \dfrac{10}{12} = \dfrac{6}{5} \times \dfrac{2 \cdot 5}{2 \cdot 6} = 1$

25. $\dfrac{5}{16} \div \dfrac{1}{8} = \dfrac{5}{2 \cdot 8} \times \dfrac{8}{1} = \dfrac{5}{2} = 2\dfrac{1}{2}$

27. $10\dfrac{3}{7} \times 5\dfrac{1}{4} = \dfrac{73}{7} \times \dfrac{21}{4} = \dfrac{73}{7} \times \dfrac{3 \cdot 7}{4}$

29. $2\dfrac{1}{8} \div \dfrac{1}{4} = \dfrac{17}{8} \times \dfrac{4}{1} = \dfrac{17}{2}$

31. $6 \times 2\dfrac{1}{2} = \dfrac{6}{1} \times$

33. $2\dfrac{1}{2} \times \dfrac{1}{10} \times \dfrac{3}{4} = \dfrac{5}{2} \times$

3

35. $2\frac{3}{4} \times 26 = \frac{11}{4} \times \frac{26}{1} = \frac{11}{2 \cdot 2} \times \frac{2 \cdot 13}{1} = \frac{143}{2} = 71\frac{1}{2}$ yards

Putting Your Skills to Work

1. $9\frac{3}{4} \times 100 = \frac{39}{4} \times \frac{100}{1} = \frac{39}{2 \cdot 2} \times \frac{2 \cdot 2 \cdot 25}{1} = \$975$

3. $59 - 56\frac{1}{4} = \frac{59}{1} - \frac{225}{4} = \frac{236}{4} - \frac{225}{4}$

$= \frac{11}{4} = 2\frac{3}{4}$

5. $5025 \div 33\frac{1}{2} = \frac{5025}{1} \div \frac{67}{2} = \frac{25 \cdot 67 \cdot 3}{1} \times \frac{2}{67}$

$= 150$

7. Answers may vary.

Exercises 0.4

1. A decimal is another way of writing a fraction whose denominator is 10, 100, 1,000 etc.

3. When multiplying 0.059 by 10,000 move the decimal point 4 places to the right.

5. $\frac{3}{20} = 20\overline{)3.00} = 0.15$
$$\begin{array}{r} 0.15 \\ 20\overline{)3.00} \\ \underline{20} \\ 100 \\ \underline{100} \end{array}$$

7. $\frac{15}{16} = 16\overline{)15.0000} = 0.9375$
$$\begin{array}{r} 0.9375 \\ 16\overline{)15.0000} \\ \underline{144} \\ 60 \\ \underline{48} \\ 120 \\ \underline{112} \\ 80 \\ \underline{80} \end{array}$$

$\frac{6}{20} = 20\overline{)6.0} = 0.3$
$$\begin{array}{r} 0.3 \\ 20\overline{)6.0} \\ \underline{60} \end{array}$$

$\frac{15}{100} = \frac{3}{20}$

$\frac{8}{00} = \frac{2}{25}$

15. $0.625 = \frac{625}{1000} = \frac{5}{8}$

17.
$$\begin{array}{r} 1.20 \\ 3.90 \\ + 2.62 \\ \hline 7.72 \end{array}$$

19.
$$\begin{array}{r} 1.0076 \\ - 0.0982 \\ \hline 0.9094 \end{array}$$

21.
$$\begin{array}{r} 22.000 \\ 0.420 \\ + 1.936 \\ \hline 24.356 \end{array}$$

23.
$$\begin{array}{r} 121.98 \\ - 34.78 \\ \hline 87.20 \end{array}$$

25.
$$\begin{array}{r} 7.12 \\ \times 2.6 \\ \hline 4272 \\ 1424 \\ \hline 18.512 \end{array}$$

27.
$$\begin{array}{r} 7.21 \\ \times 0.071 \\ \hline 721 \\ 5047 \\ \hline 0.51191 \end{array}$$

29.
$$\begin{array}{r} 0.17 \\ \times 0.0084 \\ \hline 68 \\ 000136 \\ \hline 0.001428 \end{array}$$

31.
$$\begin{array}{r} 169,000 \\ \times 0.0013 \\ \hline 507000 \\ 169000 \\ \hline 219.7000 \end{array} \text{ or } 219.7$$

33. $0.15\overline{)0.5535}$
$$\begin{array}{r} 3.69 \\ 0.15\overline{)0.5535} \\ \underline{45} \\ 103 \\ \underline{90} \\ 135 \\ \underline{135} \end{array}$$

$$
\begin{array}{r}
8.8 \\
35. \quad 3.16\overline{)27.808} \\
2528 \\
\overline{2528} \\
\underline{2528}
\end{array}
$$

$$
\begin{array}{r}
6.2 \\
37. \quad 0.005\overline{)0.0310} \\
30 \\
\overline{10} \\
\underline{10}
\end{array}
$$

$$
\begin{array}{r}
6.3 \\
39. \quad 7.4\overline{)46.62} \\
444 \\
\overline{222} \\
\underline{222}
\end{array}
$$

41. $0.76 \div 100 = 0.0076$

43. $0.02 \times 100 = 2$

45. $0.00243 \times 100,000 = 243$

47. $73,892 \div 100,000 = 0.73892$

49.
$$
\begin{array}{r}
1.936 \\
\times \ 0.003 \\
\hline
0.005808
\end{array}
$$

51.
$$
\begin{array}{r}
28.00 \\
- \quad 3.64 \\
\hline
24.36
\end{array}
$$

$$
\begin{array}{r}
317325. \\
53. \quad 0.0004\overline{)126.9300} \\
12 \\
\overline{6} \\
4 \\
\overline{29} \\
28 \\
\overline{13} \\
12 \\
\overline{10} \\
8 \\
\overline{20} \\
\underline{20}
\end{array}
$$

55. $34.72 \div 10,000 = 0.003472$

$$
\begin{array}{r}
17.85 \\
57. \quad 26.8\overline{)478.600} \\
268 \\
\overline{2106} \\
1876 \\
\overline{2300} \\
2144 \\
\overline{1560} \\
\underline{1340}
\end{array}
$$

17.9 miles per gallon

Cumulative Review Problems

59. $\dfrac{3}{8} \cdot \dfrac{12}{27} = \dfrac{3}{2 \cdot 4} \cdot \dfrac{3 \cdot 4}{3 \cdot 3 \cdot 3} = \dfrac{1}{6}$

61. $1\dfrac{3}{5} - \dfrac{1}{2} = \dfrac{8}{5} - \dfrac{1}{2} = \dfrac{16}{10} - \dfrac{5}{10} = \dfrac{11}{10} = 1\dfrac{1}{10}$

Putting Your Skills to Work

1.
Excavation	$220 \times 50 =$	11,000
Foundation	$20 \times 400 =$	8,000
Framing	$50 \times 200 =$	10,000
Roofing	$300 \times 2 =$	600
Finishing	$300 \times 16 =$	4,800
G.C. Total	$=$	34,400

3. $34,400 \div 120 = \$286.67$ per day

5. Mar 22 - Apr 1 $= 12$ days
$12 \times 286.67 \times 60\% \times 0.5 = 1032$
It will cost $1032 more.

Exercises 0.5

1. 19% means 19 out of 100 parts or $\dfrac{19}{100}$

3. $0.624 = 62.4\%$

5. $0.003 = 0.3\%$

7. $1.56 = 156\%$

9. $4\% = 0.04$

11. $0.2\% = 0.002$

13. $250\% = 2.50$

15. 35.8% of 1000
$.358 \times 1000 = 358$

17. 0.5% of 54
$0.005 \times 54 = 0.27$

19. 140% of 212
$1.40 \times 212 = 296.8$

21. 80% of 220
$0.80 \times 220 = 176$

23. What percent of 600 is 30?
$$\frac{30}{600} = \frac{1}{20} = 0.05 = 5\%$$

25. 6 is what percent of 120?
$$\frac{6}{120} = \frac{1}{20} = 0.05 = 5\%$$

27. What percent of 56 is 84?
$$\frac{84}{56} = \frac{3}{2} = 1.50 = 150\%$$

29. What percent of 500 is 2?
$$\frac{2}{500} = \frac{1}{250} = 0.004 = 0.4\%$$

31. What percent of 16 is 22?
$$\frac{22}{16} = \frac{11}{8} = 1.375 = 137.5\%$$

33. $47\% \times 30,000 = 0.47 \times 30,000$
$$= 14,100 \text{ Democrats}$$

35. $\dfrac{97}{180} = 0.53\overline{888} = 53.89\%$ smoked

37. $1.8\% \times 36,000 = 0.018 \times 36,000$
$$= 648 \text{ defective}$$

39. $2390 - 560 = 1830$ "on time"
$$\frac{1830}{2390} \approx 0.76569 \approx 77\% \text{ "on time"}$$

41. a) $65\% \times 18,600 = 12,090$ miles
b) $0.275 \times 12,090 = \$3,324.75$

Putting Your Skills to Work

1. $\dfrac{27,000}{540,000} = \dfrac{1}{20} = 0.05 = 5\%$ complete

3. $2\% \times 12,500 = 0.02 \times 12,500 = \$250$ saved

5. Total Calendar Days = 86

Total Cost = $34,400

Item	Cost Per Day	Work Complete	Percent Complete
Excavation	$ 550	$ 11,000	31.98
Concrete	800	19,000	55.23
Framing	1000	29,000	84.30
Roof	100	29,600	86.05
Finishing	120	34,400	100.00
Total	2070		

It cost $34,400 to complete in 86 calendar days at a cost per day of $400.

7. $\dfrac{29,600 + 14 \times 120}{34,400} = \dfrac{31,280}{34,400} = 0.9093 = 90.93\%$

Complete after 60 days.

Exercises 0.6

1. $878 \times 203 \approx 880 \times 200 = 176,000$

3. $58,126 \times 1786 \approx 60,000 \times 2000 = 120,000,000$

5. $93 + 87 + 56 + 22 + 34$
$$\approx 90 + 90 + 60 + 20 + 30 = 290$$

7. $23\overline{)578,962} \approx 20\overline{)600,000}$ with quotient 30,000

9. $\dfrac{0.002714}{0.0315} \approx \dfrac{0.003}{0.030} = \dfrac{1}{10} = 0.1$

11. 3.6% of $58,921.63 \approx 0.04 \times 60,000 = 2400$

13. $18\dfrac{3}{4} \times 12\dfrac{1}{8} \approx 20 \times 10 = 200$ sq. feet

15. $\dfrac{234.8}{12.6} \approx \dfrac{200}{10} \approx 20$ miles per gallon

17. $4 \times 22 \times 82 \approx 4 \times 20 \times 80 = \$6400$

19.
$$3 \times 3.95 \approx 3 \times 4.00 = 12.00$$
$$2 \times 2.59 \approx 2 \times 3.00 = 6.00$$
$$\underline{+2 \times 1.25 \approx 2 \times 1.00 = 2.00}$$
$$\$20.00$$

21. 6% of $13{,}587 \approx .06 \times 10{,}000 = \$600$

23. 8.3% of $19{,}364{,}282{,}153.18$
$$\approx 0.1 \times 20{,}000{,}000{,}000$$
$$= \$2{,}000{,}000{,}000$$

25. $43 \times 3.24 \approx 40 \times 3 = 120$
$$12 \times 120 \times 23\% \approx 10 \times 120 \times 0.2 = \$240 \text{ saved}$$

Exercises 0.7

1. Area $= 11\frac{1}{4} \times 18\frac{1}{2} = 208\frac{1}{8}$ sq. feet

$$208\frac{1}{8} \div 9 = 23\frac{1}{8} \text{ sq. yards}$$

$$\text{Cost} = 23\frac{1}{8} \times 20.00 = \$462.50$$

3. Volume $= 58\frac{1}{2} \times 18\frac{1}{2} \times \frac{1}{2} = 541\frac{1}{8}$ cu. feet

$$= \frac{541\frac{1}{8}}{27} = 20\frac{1}{24} \approx 20 \text{ cu. yards}$$

5.

Day	Jog	Walk	Rest	Walk
1	$1\frac{1}{5}$	$1\frac{3}{4}$	$2\frac{1}{2}$	1
3	$2\frac{2}{15}$	$3\frac{1}{9}$	$4\frac{4}{9}$	$1\frac{7}{9}$

Increase $1\frac{1}{3} \times 1\frac{1}{3} = \frac{16}{9}$

$$\frac{16}{9} \times 1\frac{1}{5} = \frac{16}{9} \times \frac{6}{5} = \frac{96}{45} = 2\frac{2}{15}$$

$$\frac{16}{9} \times 1\frac{3}{4} = \frac{16}{9} \times \frac{7}{4} = \frac{28}{9} = 3\frac{1}{9}$$

$$\frac{16}{9} \times 2\frac{1}{2} = \frac{16}{9} \times \frac{5}{2} = \frac{40}{9} = 4\frac{4}{9}$$

$$\frac{16}{9} \times 1 = \frac{16}{9} = 1\frac{7}{9}$$

7. Betty will have a more demanding schedule on day 3 because her second increase is based on the day 2 schedule which is greater than day 1. Melinda's second increase is still based on day 1 which is less than Betty's day 2.

9. Increase: $1\frac{1}{5} \times \frac{1}{3} = \frac{6}{5} \times \frac{1}{3} = \frac{2}{5}$

Day 7: $1\frac{1}{5} + \left(6 \times \frac{2}{5}\right) = \frac{6}{5} + \frac{12}{5} = \frac{18}{5}$
$$= 3\frac{3}{5} \text{ miles}$$

11. Total miles $= 47{,}001.2 - 45{,}678.2 = 1323$

Total gallons $= 11 + 13.2 + 12.8 + 12 = 49$

$$\frac{1323}{49} = 27 \text{ miles per gallon}$$

13. Increase 1985-1990: $10\% \times 120{,}000 = 12{,}000$

1990 Population $= 120{,}000 + 12{,}000 = 132{,}000$

Increase 1990-1995: $8\% \times 132{,}000 = 10{,}560$

1995 Population $= 132{,}000 + 10{,}560 = 142{,}560$

15. $\%$ Total (men 21 or older) $\frac{490}{1000} = 0.49 = 49\%$

$\%$ Business (men 21 or older) $\frac{380}{620} \approx 0.61 = 61\%$

$\%$ Business was greater because business travelers are more likely to be men 21 years or older.

17. Federal: $\frac{139}{1150} \approx 0.12 = 12\%$

State: $\frac{68}{1150} \approx 0.06 = 6\%$

Local: $\frac{5}{1150} \approx 0.004 = 0\%$

Total: $12\% + 6\% = 18\%$

19. $\frac{790.47}{1150} \approx 0.69 = 69\%$

Chapter 0 - Review Problems

1. $\frac{36}{48} = \frac{12 \times 3}{12 \times 4} = \frac{3}{4}$

3. $\frac{24}{72} = \frac{24 \times 1}{24 \times 3} = \frac{1}{3}$

5. $2\dfrac{5}{7} = \dfrac{(2 \times 7) + 5}{7} = \dfrac{19}{7}$

7. $\dfrac{27}{4} = 4\overline{)27} = 6\dfrac{3}{4}$
$\underline{24}$
3

9. $\dfrac{1}{3} = \dfrac{?}{45} \Rightarrow \dfrac{1 \times 15}{3 \times 15} = \dfrac{15}{45}$

11. $\dfrac{2}{5} = \dfrac{?}{55} \Rightarrow \dfrac{2 \times 11}{5 \times 11} = \dfrac{22}{55}$

13. $\dfrac{1}{12} + \dfrac{3}{8} = \dfrac{1 \times 2}{12 \times 2} + \dfrac{3 \times 3}{8 \times 3} = \dfrac{11}{24}$

15. $\dfrac{5}{12} - \dfrac{1}{16} = \dfrac{5 \times 4}{12 \times 4} - \dfrac{1 \times 3}{16 \times 3} = \dfrac{17}{48}$

17. $1\dfrac{1}{4} + 2\dfrac{7}{10} = \dfrac{5}{4} + \dfrac{27}{10} = \dfrac{25}{20} + \dfrac{54}{20} = \dfrac{79}{20} = 3\dfrac{19}{20}$

19. $3\dfrac{1}{15} - 1\dfrac{3}{20} = \dfrac{46}{15} - \dfrac{23}{20} = \dfrac{184}{60} - \dfrac{69}{60} = \dfrac{115}{60} = 1\dfrac{55}{60} = 1\dfrac{11}{12}$

21. $\dfrac{3}{5} \times \dfrac{10}{21} = \dfrac{3}{5} \times \dfrac{2 \cdot 5}{3 \cdot 7} = \dfrac{2}{7}$

23. $2\dfrac{1}{3} \times 4\dfrac{1}{2} = \dfrac{7}{3} \times \dfrac{9}{2} = \dfrac{7}{3} \times \dfrac{3 \cdot 3}{2} = \dfrac{21}{2} = 10\dfrac{1}{2}$

25. $\dfrac{4}{7} \times 5 = \dfrac{4}{7} \times \dfrac{5}{1} = \dfrac{20}{7} = 2\dfrac{6}{7}$

27. $\dfrac{1}{2} \div \dfrac{1}{8} = \dfrac{1}{2} \times \dfrac{2 \cdot 4}{1} = 4$

29. $\dfrac{5}{7} \div \dfrac{15}{14} = \dfrac{5}{7} \times \dfrac{2 \cdot 7}{3 \cdot 5} = \dfrac{2}{3}$

31. $2\dfrac{6}{7} \div \dfrac{10}{21} = \dfrac{20}{7} \times \dfrac{21}{10} = \dfrac{2 \cdot 10}{7} \times \dfrac{3 \cdot 7}{10} = 6$

33. 24.831
$\underline{-17.094}$
7.737

35. 1.600
3.210
$\underline{0.004}$
4.814

37. 100.0100
10.0010
1.1011
$\underline{1}\overline{11.1121}$

39. $362.341 \times 1000 = 362{,}341$

41. $1.08 \times 0.06 \times 160 = 10.368$

43. $71.32 \div 1000 = 0.07132$

45. $90.$
$0.015\overline{)1.350}$
$\underline{135}$
00

47. 0.5
$0.38\overline{)0.190}$
$\underline{190}$

49. $24\% = \dfrac{24}{100} = \dfrac{6}{25}$

51. $36.1\% = 0.361$

53. $125.3\% = 1.253$

55. 250% of 36
$250\% \times 36 = 2.5 \times 36 = 90$

57. What percent of 120 is 15?
$$\dfrac{15}{120} = 0.125 = 12.5\%$$

59. $36\% \times 270{,}000{,}000 = 0.36 \times 270{,}000{,}000$
$= 97{,}200{,}000$ with Type B

61. $234{,}897 \times 1{,}936{,}112 \approx 200{,}000 \times 2{,}000{,}000$
$= 400{,}000{,}000{,}000$

63. $780{,}000 - 198{,}000 \approx 800{,}000 - 200{,}000$
$= 600{,}000$

65. 18% of 56,297 $\approx 0.2 \times 60{,}000 = \$12{,}000$

67. Time: $3\dfrac{1}{2} \times 8 = \dfrac{7}{2} \times 8 = 28$ hours
$18.00 \times 28 \approx 20 \times 28 = \$560$

69. Area $= 12\dfrac{1}{2} \times 9\dfrac{2}{3} = \dfrac{25}{2} \times \dfrac{29}{3} = 120\dfrac{5}{6}$ sq. feet
$= \dfrac{120\frac{5}{6}}{9} = 13\dfrac{23}{54}$ sq. yards
Cost $= 26.00 \times 13\dfrac{23}{54} = 26 \times \dfrac{725}{54} = \$349.07$

Test - Chapter 0

1. $\dfrac{16}{18} = \dfrac{8}{9}$

3. $6\dfrac{3}{7} = \dfrac{45}{7}$

5. $\dfrac{11}{16} = \dfrac{?}{48} \Rightarrow \dfrac{11}{16} = \dfrac{33}{48}$

7. $\dfrac{1}{3} + \dfrac{1}{4} + \dfrac{5}{6} = \dfrac{8+6+20}{24} = \dfrac{34}{24} = \dfrac{17}{12} = 1\dfrac{5}{12}$

9. $1\dfrac{1}{8} + 3\dfrac{3}{4} = \dfrac{9}{8} + \dfrac{15}{4} = \dfrac{9+30}{8} = 4\dfrac{7}{8}$

11. $6 - 4\dfrac{3}{4} = \dfrac{24}{4} - \dfrac{19}{4} = \dfrac{5}{4} = 1\dfrac{1}{4}$

13. $\dfrac{3}{11} \times \dfrac{7}{5} = \dfrac{21}{55}$

15. $2\dfrac{1}{2} \times 3\dfrac{1}{4} = \dfrac{5}{2} \times \dfrac{13}{4} = \dfrac{65}{8} = 8\dfrac{1}{8}$

17. $0.72 = \dfrac{72}{100} = \dfrac{18}{25}$

19.
```
  1.60
  3.24
  9.80
 14.64
```

21.
```
 32.8
 0.04
 1.312
```

23.
```
        230.
0.056)12.880
       112
       168
       168
         0
         0
```

25. $0.073 = 7.3\%$

27. $18\% \text{ of } 350 = .18 \times 350 = 63$

29. What percent of 120 is 15?

$\dfrac{15}{120} = \dfrac{1}{8} = 0.125 = 12.5\%$

31. $\dfrac{416}{800} = \dfrac{13}{25} = 0.52 = 52\%$

33.
```
52,344)4,678,987
            100
≈ 50,000)5,000,000
         5,000,0
             000
             000
```

35.
```
  1867.85      1900
    98.87  ⇒   100
   397.49      400
 + 278.59    + 300
            $2,700
```

Practice Quiz: Sections 0.1 - 0.3

1. Change $3\dfrac{4}{5}$ to an improper fraction.

3. Add $\dfrac{3}{7} + \dfrac{5}{9}$ and write the answer in simplest form.

5. Multiply $3\dfrac{2}{5} \times 4\dfrac{2}{3}$ and write the answer in simplest form.

Practice Quiz: Sections 0.4 - 0.7

1. Change $\dfrac{2}{9}$ to a decimal.

3. Change 1.03 to a percent.

5. Estimate 312×19.

Answers to Practice Quiz

Sections 0.1 - 0.3

1. $\dfrac{19}{5}$

3. $\dfrac{62}{63}$

5. $15\dfrac{13}{15}$

Sections 0.4 - 0.7

1. $0.\overline{2}$

3. 103%

5. 6000

Pretest Chapter 1

1. $(-3)-(-6)=(-3)+(+6)=3$

3. $(-7)+(-11)=-18$

5. $(-7)(-2)(+3)(-1)=-(7)(2)(3)(1)=-42$

7. $(-7)(-4)=28$

9. $(-2)^4=16$

11. $\left(-\dfrac{2}{3}\right)^3=-\dfrac{8}{27}$

13. $-2x(3x-2xy+z)$
$=-2x(3x)+(-2x)(-2xy)+(-2x)(z)$
$=-6x^2+4x^2y-2xz$

15. $5x^2-3xy-6x^2y-8xy$
$=5x^2-6x^2y+(-3-8)xy$
$=5x^2-6x^2y-11xy$

17. $3(2x-5y)-(x-8y)$
$=6x-15y-x+8y$
$=5x-7y$

19. $6\cdot 2+8\cdot 3-4\cdot 2=12+24-8=28$

21. $-\dfrac{1}{6}\left(\dfrac{1}{2}\right)+\dfrac{3}{4}\div\dfrac{1}{12}$
$=-\dfrac{1}{12}+9=-\dfrac{1}{12}+\dfrac{108}{12}=\dfrac{107}{12}=8\dfrac{11}{12}$

23. If $x=-2$, then $3x^2-5x-4$
$=3(-2)^2-5(-2)-4=12+10-4=18$

25. If $F=77$, then
$C=\dfrac{5}{9}(F-32)=\dfrac{5}{9}[77-32]=\dfrac{5}{9}(45)=25^0C$

27. $2x^2-3x[2x-(x+2y)]$
$=2x^2-3x(2x-x-2y)$
$=2x^2-3x(x-2y)$
$=2x^2-3x^2+6xy$
$=-x^2+6xy$

Exercises 1.1

	Number	Rational Number	Irrational Number
1.	23	x	
3.	π		x
5.	-6.666...	x	
7.	-2.3434...	x	
9.	$\sqrt{2}$		x

11. A check for \$53 $\Rightarrow -53$

13. Down 85 feet $\Rightarrow -85$

15. Rise 7^0 F $\Rightarrow +7$

17. Opposite of $\dfrac{3}{4}$ is $-\dfrac{3}{4}$

19. Opposite of $-\dfrac{5}{13}$ is $+\dfrac{5}{13}$

21. $(-9)+(+5)=-4$

23. $(-8)+(-5)=-13$

25. $\left(-\dfrac{1}{3}\right)+\left(+\dfrac{2}{3}\right)=\dfrac{1}{3}$

27. $(+35)+(+20)=55$

29. $(-15)+(-16)=-31$

31. $\left(-\dfrac{2}{13}\right)+\left(-\dfrac{5}{13}\right)=-\dfrac{7}{13}$

33. $(-1.5)+(-2.3)=-3.8$

35. $(+0.6)+(-0.2)=0.4$

37. $(-12)+(-13)=-25$

39. $\left(+\dfrac{1}{3}\right)+\left(-\dfrac{1}{4}\right)=\left(+\dfrac{4}{12}\right)+\left(-\dfrac{3}{12}\right)=\dfrac{1}{12}$

41. $(+2)+(-7)+(-6)=(-5)+(-6)=-11$

43. $(-3)+(+8)+(+5)+(-7)=(+5)+(+5)+(-7)$
$=(10)+(-7)=3$

45. $(-2)+(+8)+(-3)+(-5)=(+6)+(-3)+(-5)$
$=(+3)+(-5)=-2$

47. $(+31)+(-16)+(+15)+(-17)$
$=(+15)+(+15)+(-17)=(+30)+(-17)=13$

49. $(+0.5)+(-3.2)+(-2)+(+1.5)$
$=(-2.7)+(-2)+(1.5)=(-4.7)+(1.5)$
$=-3.2$

51. $21+(-4)=17$

53. $-98+103=5$

55. $-\dfrac{7}{8}+\dfrac{1}{2}=-\dfrac{7}{8}+\dfrac{4}{8}=-\dfrac{3}{8}$

57. $-\dfrac{2}{3}+\left(-\dfrac{1}{4}\right)=-\dfrac{8}{12}+\left(-\dfrac{3}{12}\right)=-\dfrac{11}{12}$

59. $5.7+(-9.1)=-3.4$

61. $-14+(-23)+5=-37+5=-32$

63. $18+(-39)+25+(-3)$
$=-21+25+(-3)$
$=4+(-3)=1$

65. $12.59+1.98+(-0.84)$
$=14.57+(-0.84)=13.73$

67. $-18+12=-6^{0}\text{F}:$ today's high

69. $-85+(-180)=-265$ feet

71. $75+(-18)=57$ degrees at 5 pm

73. $-258+(-32)+150$
$=-290+150=-140$
He owes $140

75. $-18+?=10$
$-18+28=10$

77. Under one of the following cases:
   a) $|x|>|y|$ with x negative and y positive
   b) $|y|>|x|$ with x positive and y negative
   c) x and y both negative

79. $50+(-89)=-39$
Glen needs at least $39.00

Putting Your Skills to Work

1. $334.56+46.78+(-12.39)+(-67.99)+(-35)$
$=381.34 \qquad + \qquad (-115.38)$
$=\$265.96$ in her checkbook

3. $(-125.38)+(-72.50)+(-98.12)=-296$
$-296+268=-28$
$-28.00+50.25=22.25$
The outstanding checks total $296, not $268. He may have also neglected to add the deposit of $50.25.

Exercises 1.2

1. $(+8)-(+5)=(+8)+(-5)=3$

3. $(+7)-(-3)=(+7)+(+3)=10$

5. $(-12)-(-4)=(-12)+(+4)=-8$

7. $(+20)-(+46)=(+20)+(-46)=-26$

9. $(-32)-(-6)=(-32)+(+6)=-26$

11. $(-52)-(-60)=(-52)+(+60)=8$

13. $(0)-(-5)=(0)+(+5)=5$

15. $(+15)-(+20)=(+15)+(-20)=-5$

17. $(-18)-(-18)=(-18)+(+18)=0$

19. $(-17)-(-13)=(-17)+(+13)=-4$

21. $(-0.6)-(+0.3)=(-0.6)+(-0.3)=-0.9$

23. $(+2.64)-(-1.83)=(+2.64)+(+1.83)=4.47$

25. $(-1.5)-(-3.5)=(-1.5)+(+3.5)=2.0$

27. $\left(+\dfrac{3}{4}\right)-\left(-\dfrac{3}{5}\right)=\left(+\dfrac{15}{20}\right)+\left(+\dfrac{12}{20}\right)=\dfrac{27}{20}=1\dfrac{7}{20}$

29. $\left(-\dfrac{1}{5}\right)-\left(+\dfrac{3}{8}\right)=\left(-\dfrac{8}{40}\right)+\left(-\dfrac{15}{40}\right)=-\dfrac{23}{40}$

31. $34 - 87 = 34 + (-87) = -53$

33. $-67 - 32 = (-67) + (-32) = -99$

35. $2.3 - (-4.8) = 2.3 + (+4.8) = 7.1$

37. $3 - \dfrac{1}{5} = \dfrac{15}{5} + \left(-\dfrac{1}{5}\right) = \dfrac{14}{5} = 2\dfrac{4}{5}$

39. $\dfrac{4}{9} - (-14) = \dfrac{4}{9} + \left(+\dfrac{126}{9}\right) = \dfrac{130}{9} = 14\dfrac{4}{9}$

41. $-\dfrac{3}{10} - \dfrac{3}{4} = \left(-\dfrac{6}{20}\right) + \left(-\dfrac{15}{20}\right) = -\dfrac{21}{20} = -1\dfrac{1}{20}$

43. $-135 - (-126.5) = -135 + (+126.5) = -8.5$

45. $0.0067 - (-0.0432)$
    $= 0.0067 + (+0.0432) = 0.0499$

47. $\dfrac{1}{5} - 6 = \dfrac{1}{5} + \left(-\dfrac{30}{5}\right) = -\dfrac{29}{5} = -5\dfrac{4}{5}$

49. $-\dfrac{3}{5} - \left(-\dfrac{3}{10}\right) = -\dfrac{6}{10} + \left(+\dfrac{3}{10}\right) = -\dfrac{3}{10}$

51. $-2 - (-9) = -2 + (+9) = 7$

53. $-35 - (13) = -35 + (-13) = -48$

55. $7 + (-6) - (+3) = 7 + (-6) + (-3)$
    $= 1 + (-3) = -2$

57. $-10 + 6 - (-15) = -10 + 6 + (+15)$
    $= -4 + (+15) = 11$

59. $7 + (-42) - 27 = 7 + (-42) + (-27)$
    $= -35 + (-27) = -62$

61. $-31 - (-14) - 17 = -31 + (+14) + (-17)$
    $= -17 + (-17) = -34$

63. $-84 + 12 - (-45) - (+14)$
    $= -84 + 12 + (+45) + (-14)$
    $= -72 + (+45) + (-14)$
    $= -27 + (-14) = -41$

65. $42 - (-30) - 65 - (-11) + 20$
    $= 42 + (+30) + (-65) + (+11) + 20$
    $= 72 + (-65) + (+11) + 20$
    $= 7 + (+11) + 20$
    $= 18 + 20 = 38$

67. $112 - (-37) = 112 + (+37) = \$149$

69. $-45 - (-29) = -45 + (+29)$
    $= -16^{0}\,\text{F}$

Cumulative Review Problems

71. $-37 + (-14) = -51$

73. $-21 + 13 = -8^{0}\,\text{C}$

Exercises 1.3

1. $(3)(-12) = -36$

3. $(-6)(-5) = 30$

5. $(-5)(-12) = 60$

7. $-8 \times 3 = -24$

9. $0(-12) = 0$

11. $16 \times 1.5 = 24$

13. $(-1.32)(-0.2) = 0.264$

15. $(-1.8)(-0.03) = 0.054$

17. $\left(\dfrac{12}{5}\right)(-10) = \left(\dfrac{12}{5}\right)\left(-\dfrac{10}{1}\right) = -24$

19. $\left(-\dfrac{3}{8}\right)\left(-\dfrac{5}{6}\right) = \left(-\dfrac{3}{8}\right)\left(-\dfrac{5}{2 \cdot 3}\right) = \dfrac{5}{16}$

21. $(12) \div (-4) = \dfrac{12}{-4} = -3$

23. $0 \div (-15) = 0$

25. $-42 \div 7 = -6$

27. $-360 \div (-10) = \dfrac{-360}{-10} = 36$

29. $105 \div (-7) = \dfrac{105}{-7} = -15$

31. $(-0.6) \div 0.3 = \dfrac{-0.6}{0.3} = -2$

33. $(-2.4) \div (-0.6) =$

$\qquad \dfrac{-2.4}{-0.6} = 4$

35. $(-7.2) \div 8 = \dfrac{-7.2}{8} = -0.9$

37. $\left(-\dfrac{1}{5}\right) \div \left(\dfrac{2}{3}\right) = \left(-\dfrac{1}{5}\right)\left(\dfrac{3}{2}\right) = -\dfrac{3}{10}$

39. $\left(-\dfrac{3}{4}\right) \div \left(-\dfrac{7}{12}\right) = \left(-\dfrac{3}{4}\right)\left(-\dfrac{4 \cdot 3}{7}\right) = \dfrac{9}{7} = 1\dfrac{2}{7}$

41. $\dfrac{12}{-\frac{2}{5}} = \left(\dfrac{12}{1}\right)\left(-\dfrac{5}{2}\right) = -30$

43. $\dfrac{-\frac{3}{8}}{-\frac{2}{3}} = \left(-\dfrac{3}{8}\right)\left(-\dfrac{3}{2}\right) = \dfrac{9}{16}$

45. $\dfrac{\frac{2}{5}}{-\frac{8}{15}} = \left(\dfrac{2}{5}\right)\left(-\dfrac{3 \cdot 5}{2 \cdot 4}\right) = -\dfrac{3}{4}$

47. $(-1)(-2)(-3)(4) = -(1)(2)(3)(4) = -24$

49. $(-1)(-3)(-2)(-2)(3) = +(1)(3)(2)(2)(3) = 36$

51. $(-3)(-2)\left(\dfrac{1}{3}\right)(-4)(2) = -16$

53. $(-3)(-0.03)(100)(-2) = -18$

55. $\left(-\dfrac{5}{12}\right)\left(-\dfrac{4}{9}\right)\left(-\dfrac{3}{5}\right) = -\left(\dfrac{5}{3 \cdot 4}\right)\left(\dfrac{4}{9}\right)\left(\dfrac{3}{5}\right) = -\dfrac{1}{9}$

57. $\left(-\dfrac{3}{4}\right)\left(-\dfrac{7}{15}\right)\left(-\dfrac{8}{21}\right)\left(-\dfrac{5}{9}\right)$

$\qquad = +\left(\dfrac{3}{4}\right)\left(\dfrac{7}{3 \cdot 5}\right)\left(\dfrac{4 \cdot 2}{3 \cdot 7}\right)\left(\dfrac{5}{9}\right) = \dfrac{2}{27}$

Mixed Review

59. $(-20) \div 5 = \dfrac{-20}{5} = -4$

61. $5 + (-7) = -2$

63. $8 - (-9) = 8 + (9) = 17$

65. $6(-12) = -72$

67. $(-37) \div 37 = \dfrac{-37}{37} = -1$

Applications

69. $\ ? - 100 = -70$

$\quad 30 - 100 = -70$

$\quad$ \$30 in the account

71. $r = -30$ meters per second

$\quad t = -5$ seconds

$\quad d = rt = (-30)(-5) = 150$

$\quad d = 150$ meters to the right of zero.

Cumulative Review Problems

73. $(-13) + (-39) + (-20) = (-52) + (-20) = -72$

75. $(-37) - (51) = (-37) + (-51) = -88$

Exercises 1.4

1. $4^3 = (4)(4)(4) = 64$

3. $(-5)^4 = (-5)(-5)(-5)(-5) = 625$

5. $(-7)^3 = (-7)(-7)(-7) = -343$

7. $2^3 = 8$

9. $3^4 = 81$

11. $7^3 = 343$

13. $(-3)^3 = -27$

15. $(-2)^6 = 64$

17. $(-6)^3 = -216$

19. $\left(\dfrac{1}{4}\right)^2 = \dfrac{1}{16}$

14

21. $\left(\dfrac{2}{5}\right)^3 = \dfrac{8}{125}$

23. $(0.3)^2 = 0.09$

25. $(0.5)^3 = 0.125$

27. $(-8)^4 = 4096$

29. $-8^4 = -4096$

31. $(-0.6)^5 = -0.07776$

33. $(0.1)^6 = 0.000001$

35. $(-5)^2(3) = (25)(3) = 75$

37. $2^3(3)^2(-10)^3 = (8)(9)(-1000) = -72,000$

39. $(-7)4^2(3)^2 = (-7)(16)(9) = -1008$

41. $\left(\dfrac{2}{3}\right)^2\left(\dfrac{1}{4}\right)3^3 = \left(\dfrac{4}{9}\right)\left(\dfrac{1}{4}\right)\left(\dfrac{27}{1}\right) = 3$

43. $3^3 - 2^4 = 27 - 16 = 11$

45. $-7 - (-3)^3 = -7 - (-27) = -7 + 27 = 20$

To Think About

47. $(?)^6 = 64, (-2)^6 = 64, (2)^6 = 64, -2 \text{ and } 2$

Cumulative Review Problems

49. $\dfrac{3}{4} \div \left(-\dfrac{9}{20}\right) = \left(\dfrac{3}{4}\right)\left(-\dfrac{20}{9}\right) = \left(\dfrac{3}{4}\right)\left(-\dfrac{4 \cdot 5}{3 \cdot 3}\right) = -\dfrac{5}{3} = -1\dfrac{2}{3}$

51. $(-2.1)(-1.2) = 2.52$

Exercises 1.5

1. A <u>variable</u> is a symbol used to represent an unknown number.

3. $3x^2yx^6$ is a <u>monomial</u>.

5. $a(b - c) = ab - ac$ works:
$$a(b - c) = a[b + (-c)]$$
$$= ab + (a)(-c)$$
$$= ab + (-ac)$$
$$= ab - ac$$

7. $x(2x - y) = (x)(2x) - xy$
$$= 2x^2 - xy$$

9. $-2(3x^2 - 2) = (-2)(3x^2) + (-2)(-2)$
$$= -6x^2 + 4$$

11. $5a(-3a - 2b^2) = 5a(-3a) - 5a(2b^2)$
$$= -15a^2 - 10ab^2$$

13. $-3x(x + 2y - 1) = (-3x)(x) + (-3x)(2y) + (-3x)(-1)$
$$= -3x^2 - 6xy + 3x$$

15. $-(5x^2y - 2xy^2 - xy)$
$$= (-1)(5x^2y) + (-1)(-2xy^2) + (-1)(-xy)$$
$$= -5x^2y + 2xy^2 + xy$$

17. $-5(3x + 9 - 7y) = (-5)(3x) + (-5)(9) + (-5)(-7y)$
$$= -15x - 45 + 35y$$

19. $\dfrac{1}{4}(x^2 + 2x - 8) = \dfrac{1}{4}x^2 + \left(\dfrac{1}{4}\right)(2x) + \left(\dfrac{1}{4}\right)(-8)$
$$= \dfrac{x^2}{4} + \dfrac{x}{2} - 2$$

21. $\dfrac{4}{7}(7x^2 - 21x - 3) = \dfrac{4}{7}(7x^2) + \dfrac{4}{7}(-21x) + \dfrac{4}{7}(-3)$
$$= 4x^2 - 12x - \dfrac{12}{7}$$

23. $\dfrac{y}{3}(3y - 4x - 6) = \dfrac{y}{3}(3y) + \dfrac{y}{3}(-4x) - \dfrac{y}{3}(6)$
$$= y^2 - \dfrac{4xy}{3} - 2y$$

25. $3a(2a + b - c - 4) = 3a(2a) + 3ab + 3a(-c) + 3a(-4)$
$$= 6a^2 + 3ab - 3ac - 12a$$

27. $(4 - 2x)(-3) = 4(-3) + (-2x)(-3) = -12 + 6x$

29. $(-5x^3 + x)(-5) = -5x^3(-5) + x(-5) = 25x^3 - 5x$

31. $3x(4x - 5y - 6) = 3x(4x) + 3x(-5y) + 3x(-6)$
$$= 12x^2 - 15xy - 18x$$

33. $-4x(x + y - 5) = -4x^2 + (-4x)y + (-4x)(-5)$
$$= -4x^2 - 4xy + 20x$$

35. $-ab(4a - 2b - 1)$
$$= (-ab)(4a) + (-ab)(-2b) + (-ab)(-1)$$
$$= -4a^2b + 2ab^2 + ab$$

37. $4xy(-2x + y - 3) = 4xy(-2x) + 4xy(y) + 4xy(-3)$
$$= -8x^2y + 4xy^2 - 12xy$$

39. $2.5(1.5a^2 - 3.5a + 2.0)$
$$= (2.5)(1.5a^2) + (2.5)(-3.5a) + (2.5)(2.0)$$
$$= 3.75a^2 - 8.75a + 5$$

41. $-0.4a(0.3a - 0.2b + 0.02b^2)$
$$= (-0.4a)(0.3a) + (-0.4a)(-0.2b) + (-0.4a)(0.02b^2)$$
$$= -0.12a^2 + 0.08ab - 0.008ab^2$$

Cumulative Review Problems

43. $-18 + (-20) + 36 + (-14)$
$$= -38 + 36 + (-14)$$
$$= -2 + (-14) = -16$$

45. $(-1)^{200} = 1$

Exercises 1.6

1. 5 miles + 7 inches + 3 miles
   Like terms: 5 miles and 3 miles

3. $5a - 2b - 12a$
   Like terms: $5a$ and $-12a$

5. $3xy - 5x + 6x + 2xy$
   Like terms: $3xy$ and $2xy$
   $\qquad\qquad -5x$ and $6x$

7. $7x + 3x = (7 + 3)x = 10x$

9. $-12x^3 - 16x^3 = (-12 - 16)x^3 = -28x^3$

11. $10x^4 + 8x^4 + 7x^2 = (10 + 8)x^4 + 7x^2$
$$= 18x^4 + 7x^2$$

13. $5a + 2b - 7a^2$
   There are no like terms so it can't be simplified.

15. $2ab + 1 - 6ab - 8$
$$= (2 - 6)ab + 1 - 8 = -4ab - 7$$

17. $1.3x - 2.6y + 5.8x - 0.9y$
$$= (1.3 + 5.8)x + (-2.6 - 0.9)y = 7.1x - 3.5y$$

19. $1.6x - 2.8y - 3.6x - 5.9y$
$$= (1.6 - 3.6)x + (-2.8 - 5.9)y = -2x - 8.7y$$

21. $\dfrac{1}{2}x^2 - 3y - \dfrac{1}{3}y + \dfrac{1}{4}x^2$
$$= \left(\dfrac{1}{2} + \dfrac{1}{4}\right)x^2 + \left(-3 - \dfrac{1}{3}\right)y = \dfrac{3}{4}x^2 - \dfrac{10}{3}y$$

23. $\dfrac{1}{2}x - \dfrac{1}{4}y - \dfrac{3}{4}x - \dfrac{1}{8}y$
$$= \left(\dfrac{1}{2} - \dfrac{3}{4}\right)x + \left(-\dfrac{1}{4} - \dfrac{1}{8}\right)y = -\dfrac{1}{4}x - \dfrac{3}{8}y$$

25. $3x + 2y - 6 - 8x - 9y - 14$
$$= (3 - 8)x + (2 - 9)y - 6 - 14 = -5x - 7y - 20$$

27. $5x^2y - 10xy^6 + 6xy^2 - 7xy^2$
$$= 5x^2y - 10xy^6 + (6 - 7)xy^2$$
$$= 5x^2y - 10xy^6 - xy^2$$

29. $2ab + 5bc - 6ac - 2ab$
$$= (2 - 2)ab + 5bc - 6ac$$
$$= 5bc - 6ac$$

31. $x^3 - 3x^2 + 1 - 2x + 6x^2 - x^3$
$$= (1 - 1)x^3 + (-3 + 6)x^2 + 1 - 2x$$
$$= 3x^2 - 2x + 1$$

33. $2y^2 - 8y + 9 - 12y^2 - 8y + 3$
$$= (2 - 12)y^2 + (-8 - 8)y + 9 + 3$$
$$= -10y^2 - 16y + 12$$

35. $ab + 3a - 4ab + 2a - 8b$

$= (1-4)ab + (3+2)a - 8b$

$= -3ab + 5a - 8b$

37. $3(x+y) - 5(-2y + 3x)$

$= 3x + 3y + 10y - 15x$

$= -12x + 13y$

39. $5a(a + 3b) - 2(a^2 - 6ab)$

$= 5a^2 + 15ab - 2a^2 + 12ab$

$= 3a^2 + 27ab$

41. $-7(a^2 + 3b) + 4(2b - 8a^2)$

$= -7a^2 - 21b + 8b - 32a^2$

$= -39a^2 - 13b$

43. $7(3-x) - 6(8 - 13x)$

$= 21 - 7x - 48 + 78x$

$= 71x - 27$

Cumulative Review Problems

45. $\left(-\dfrac{5}{3}\right)\left(\dfrac{1}{2}\right) = -\dfrac{5}{6}$

47. $\left(\dfrac{5}{7}\right) \div \left(-\dfrac{14}{3}\right) = \left(\dfrac{5}{7}\right)\left(-\dfrac{3}{14}\right) = -\dfrac{15}{98}$

Exercises 1.7

Use the expression "three twos plus four fives"
in Exercises 1 & 3.

1. $3(2) + 4(5)$

3. a) $3(2) + 4(5) \Rightarrow 6 + 4(5) \Rightarrow 10(5) = \$50$

   b) $3(2) + 4(5) = 6 + 20 = \$26$

5. $4^2 + 2(4) = 16 + 2(4) = 16 + 8 = 24$

7. $5 + 6 \cdot 2 \div 4 - 1 = 5 + 12 \div 4 - 1$

$= 5 + 3 - 1 = 7$

9. $(3-5)^2 \cdot 6 \div 4 = (-2)^2 \cdot 6 \div 4$

$= 4 \cdot 6 \div 4 = 24 \div 4 = 6$

11. $8 - 2^3 \cdot 5 + 3 = 8 - 8 \cdot 5 + 3 = 8 - 40 + 3 = -29$

13. $4 + 27 \div 3 \cdot 2 - 8 = 4 + 9 \cdot 2 - 8$

$= 4 + 18 - 8 = 14$

15. $5(3)^3 - 20 \div (-2) = 5(27) - 20 \div (-2)$

$= 135 + 10 = 145$

17. $3(5-7)^2 - 6(3) = 3(-2)^2 - 6(3)$

$= 3(4) - 6(3) = 12 - 18 = -6$

19. $5 \times 6 - (3-5)^2 + 8 \cdot 2 = 5 \times 6 - (-2)^2 + 8 \cdot 2$

$= 5 \times 6 - 4 + 8 \cdot 2$

$= 30 - 4 + 16 = 42$

21. $\left(\dfrac{1}{2}\right) \div \left(\dfrac{2}{3}\right) + 6 \cdot \left(\dfrac{1}{4}\right)$

$= \left(\dfrac{3}{4}\right) + \left(\dfrac{6}{4}\right) = \dfrac{9}{4} = 2\dfrac{1}{4}$

23. $0.8 + 0.3(0.6 - 0.2)^2$

$= 0.8 + 0.3(0.4)^2$

$= 0.8 + 0.3(.16)$

$= 0.8 + 0.048 = 0.848$

25. $\dfrac{3}{4}\left(-\dfrac{2}{5}\right) - \left(-\dfrac{3}{5}\right) = -\dfrac{3}{10} + \dfrac{6}{10} = \dfrac{3}{10}$

27. $7.2 - (-3.8) + (5.5)^2$

$= 7.2 - (-3.8) + 30.25$

$= 7.2 + 3.8 + 30.25 = 41.25$

29. $\left(\dfrac{1}{2}\right)^3 + \left(\dfrac{1}{4}\right) - \left(\dfrac{1}{6} - \dfrac{1}{12}\right) - \dfrac{2}{3} \cdot \left(\dfrac{1}{4}\right)^2$

$= \left(\dfrac{1}{2}\right)^3 + \dfrac{1}{4} - \left(\dfrac{1}{12}\right) - \dfrac{2}{3} \cdot \left(\dfrac{1}{4}\right)^2$

$= \dfrac{1}{8} + \dfrac{1}{4} - \dfrac{1}{12} - \dfrac{2}{3} \cdot \dfrac{1}{16}$

$= \dfrac{1}{8} + \dfrac{1}{4} - \dfrac{1}{12} - \dfrac{1}{24}$

$= \dfrac{6}{24} = \dfrac{1}{4}$

Cumulative Review Problems

31. $(2x - 3y) - (x - 3y)$

$= 2x - 3y - x + 3y = x$

17

33. $-1^{2086} = -1$

Exercises 1.8

1. If $x = 3$, then $-2x + 1$
$= -2(3) + 1 = -6 + 1 = -5$

3. If $x = -8$, then $\frac{1}{2}x + 2$
$= \frac{1}{2}(-8) + 2 = -4 + 2 = -2$

5. If $x = \frac{1}{2}$, then $5x + 10$
$= 5\left(\frac{1}{2}\right) + 10 = \frac{5}{2} + 10 = \frac{25}{2} = 12\frac{1}{2}$

7. If $x = 7$, then $2 - 4x$
$= 2 - 4(7) = 2 - 28 = -26$

9. If $x = 3$, then $x^2 + 2x$
$= (3)^2 + 2(3) = 9 + 6 = 15$

11. If $x = -1$, then $3x^2 = 3(-1)^2 = 3$

13. If $x = 5$, then $-2x^2 = -2(5)^2 = -50$

15. If $x = -2$, then $9x + 13$
$= 9(-2) + 13 = -18 + 13 = -5$

17. If $x = -3$, then $2x^2 + 3x$
$= 2(-3)^2 + 3(-3) = 18 - 9 = 9$

19. If $x = 3$, then $(2x)^2 + x$
$= [2(3)]^2 + 3 = 36 + 3 = 39$

21. If $x = -2$, then $2 - (-x)^2$
$= 2 - [-(-2)]^2 = 2 - (2)^2 = 2 - 4 = -2$

23. If $x = -2$, then $16 - 7x + x^2$
$= 16 - 7(-2) + (-2)^2 = 16 + 14 + 4 = 34$

25. If $x = 2$, then $2x^2 + 3x + 6$
$= 2(2)^2 + 3(2) + 6 = 8 + 6 + 6 = 20$

27. If $x = -4$, then $\frac{1}{2}x^2 - 3x + 9$
$= \frac{1}{2}(-4)^2 - 3(-4) + 9 = 8 + 12 + 9 = 29$

29. (a) If $x = 4$, then $x^2 - 3x + 4$
$= 4^2 - 3(4) + 4 = 8$

(b) If $x = 0$, then $x^2 - 3x + 4$
$= 0^2 - 3(0) + 4 = 4$

(c) If $x = -2$, then $x^2 - 3x + 4$
$= (-2)^2 - 3(-2) + 4 = 14$

31. (a) If $x = 2$, then $6 - 2x - 3x^2$
$= 6 - 2(2) - 3(2)^2 = -10$

(b) If $x = 0$, then $6 - 2x - 3x^2$
$= 6 - 2(0) - 3(0)^2 = 6$

(c) If $x = -1$, then $6 - 2x - 3x^2$
$= 6 - 2(-1) - 3(-1)^2 = 5$

33. If $x = 4$ and $y = -1$, then $2x^2 - 3xy + 2y$
$= 2(4)^2 - 3(4)(-1) + 2(-1) = 32 + 12 - 2 = 42$

35. If $a = 3$, $b = 2$, and $c = -4$, then
$a^2 - 2ab + 2c^2 = 3^2 - 2(3)(2) + 2(-4)^2$
$= 9 - 12 + 32 = 29$

Applications

37. $A = \frac{1}{2}bh$
$= \frac{1}{2}(13)(14)$
$= 91$ sq. in.

14 in
13 in

39. $A = \pi r^2$
$= (3.14)(5)^2$
$= 78.5$ sq. m

5 m

41. If C is -10^0C, then
$F = \frac{9}{5}C + 32 = \frac{9}{5}(-10) + 32 = 14^0\text{F}$

18

43. If $r = 7\ m$, $A = \pi r^2 = (3.14)(7)^2$
$$= 153.86 \text{ sq. m}$$
$$\text{Cost} = (153.86 \text{ sq. m})(\$350/\text{sq. m})$$
$$= \$53,851.00$$

45. When $C = -60^0$, $F = \dfrac{9}{5}C + 32 = \dfrac{9}{5}(-60) + 32$
$$= -76^0$$
When $C = -30^0$, $F = \dfrac{9}{5}(-30^0) + 32$
$$= -22^0$$
The range is $-76^0\,$F to $-22^0\,$F.

47.
$$K = 1.61r$$
When $K = 20\ km$,
$$r = \frac{20}{1.61} = 12.4 \text{ miles}$$

Cumulative Review Problems

49. $3(x - 2y) - (x^2 - y) - (x - y)$
$$= 3x - 6y - x^2 + y - x + y$$
$$= -x^2 + 2x - 4y$$

51. $(-0.3) \times (2.4) = -0.72$

Putting Your Skills to Work

1. If $n = 32$, and $S = 1575$, then
$$R = \frac{3}{5}S\left(1 + \frac{2n}{100}\right) = \frac{3}{5}(1575)\left(1 + \frac{2 \cdot 32}{100}\right)$$
$$= 945\left(1 + \frac{64}{100}\right) = 945\left(\frac{164}{100}\right) = 1549.8$$

Monthly retirement income is \$1549.80

3. Present position:
$$R = \frac{3}{5}S\left(1 + \frac{2n}{100}\right)$$
$$= \frac{3}{5} \cdot 1700\left(1 + \frac{2.25}{100}\right)$$
$$= 1020\left(1 + \frac{1}{2}\right)$$
$$= 1020\left(\frac{3}{2}\right) = 1530$$

With training:
$$R = \frac{3}{5}S\left(1 + \frac{2n}{100}\right)$$
$$= \frac{3}{5}(2100)\left(1 + \frac{2 \cdot 20}{100}\right)$$
$$= 1260\left(1 + \frac{2}{5}\right)$$
$$= 1260\left(\frac{7}{5}\right) = 1764$$

Additional monthly income:
$$1764 - 1530 = \$234$$

Exercises 1.9

1. $-3x - 2y = -(3x + 2y)$

3. To simplify expressions with grouping symbols, we use the <u>distributive</u> property.

5. $x - 4(x + y) = x - 4x - 4y = -3x - 4y$

7. $2(a + 3b) - 3(b - a)$
$$= 2a + 6b - 3b + 3a$$
$$= 5a + 3b$$

9. $x^2(x^2 - 3y^2) - 2(x^4 - x^2y^2)$
$$= x^4 - 3x^2y^2 - 2x^4 + 2x^2y^2$$
$$= -x^4 - x^2y^2$$

11. $5[3 + 2(x - 26) + 3x]$
$$= 5[3 + 2x - 52 + 3x]$$
$$= 5(5x - 49)$$
$$= 25x - 245$$

13. $2x + \left[4x^2 - 2(x-3)\right]$

$= 2x + \left(4x^2 - 2x + 6\right)$

$= 4x^2 + 6$

15. $4y\left[-3y^2 + 2(4-y)\right]$

$= 4y\left(-3y^2 + 8 - 2y\right)$

$= -12y^3 - 8y^2 + 32y$

17. $2(x-2y) - \left[3 - 2(x-y)\right]$

$= 2x - 4y - (3 - 2x + 2y)$

$= 2x - 4y - 3 + 2x - 2y$

$= 4x - 6y - 3$

19. $2\left[a - 3b(a+b) - b^2\right]$

$= 2\left(a - 3ab - 3b^2 - b^2\right)$

$= 2\left(a - 3ab - 4b^2\right)$

$= 2a - 6ab - 8b^2$

21. $x\left(x^2 + 2x - 3\right) - 2\left(x^3 + 6\right)$

$= x^3 + 2x^2 - 3x - 2x^3 - 12$

$= -x^3 + 2x^2 - 3x - 12$

23. $3a^2 - 4\left[2b - 3b(b+2)\right]$

$= 3a^2 - 4\left(2b - 3b^2 - 6b\right)$

$= 3a^2 - 4\left(-4b - 3b^2\right)$

$= 3a^2 + 16b + 12b^2$

25. $5x - \left\{4y - 2\left[2x + 3(x-y)\right]\right\}$

$= 5x - \left[4y - 2(2x + 3x - 3y)\right]$

$= 5x - \left[4y - 2(5x - 3y)\right]$

$= 5x - (4y - 10x + 6y)$

$= 5x - (-10x + 10y)$

$= 5x + 10x - 10y$

$= 15x - 10y$

27. $-\left\{4a^2 - 2\left[3a^2 - (b - a^2)\right]\right\}$

$= -\left[4a^2 - 2(3a^2 - b + a^2)\right]$

$= -\left[4a^2 - 2(4a^2 - b)\right]$

$= -\left(4a^2 - 8a^2 + 2b\right)$

$= -\left(-4a^2 + 2b\right)$

$= 4a^2 - 2b$

29. $-4\left\{3a^2 - 2\left[4a^2 - (b + a^2)\right]\right\}$

$= -4\left[3a^2 - 2(4a^2 - b - a^2)\right]$

$= -4\left[3a^2 - 2(3a^2 - b)\right]$

$= -4\left(3a^2 - 6a^2 + 2b\right)$

$= -4\left(-3a^2 + 2b\right)$

$= 12a^2 - 8b$

Cumulative Review Problems

31. If $C = 36.4$, then $F = 1.8C + 32$

$F = 1.8C + 32 = 1.8(36.4) + 32 = 97.52^0\,F.$

33. $3 - 4(-2)^3 = 3 - 4(-8) = 3 + 32 = 35$

Putting Your Skills To Work

Use $P = n(1+r)^t$ in Exercises 1 and 3.

1. (a) If $n = 7000$, $r = 12\%$, and $t = 8$.

$P = 7000(1 + 0.12)^8 = 17{,}332$

(b) If $n = 17{,}332$, $r = 8\%$, and $t = 5$.

$P = 17{,}332(1 + 0.08)^5 = 25{,}466$

(c) If $n = 25{,}466$, $r = 4\%$, and $t = 3$

$P = 25{,}466(1 + 0.04)^3 = 28{,}646$

The population will be 28,646

3. Crestwood City: $n = 45,000$

   $r = 4\%, \ t = 6$

   $P = 45,000(1 + 0.04)^6 = 56,939$

   Millville: $n = 38,000, \ r = 8\%, \ t = 6$

   $P = 38,000(1 + 0.08)^6 = 60,301$

   Millville will be larger by

   $60,301 - 56,939 = 3,362$ people

Chapter 1 - Review Problems

1. $(-6) + (-2) = -8$

3. $(+5) + (-2) + (-12) = (+3) + (-12) = -9$

5. $\left(+\dfrac{1}{2}\right) + \left(-\dfrac{5}{6}\right) = \left(+\dfrac{3}{6}\right) + \left(-\dfrac{5}{6}\right) = -\dfrac{2}{6} = -\dfrac{1}{3}$

7. $\left(+\dfrac{3}{4}\right) + \left(-\dfrac{1}{12}\right) + \left(-\dfrac{1}{2}\right)$

   $= \left(+\dfrac{9}{12}\right) + \left(-\dfrac{1}{12}\right) + \left(-\dfrac{6}{12}\right) = \dfrac{2}{12} = \dfrac{1}{6}$

9. $(-5) + (-15) + (+6) + (-9) + (+10)$

   $= (-20) + (+6) + (-9) + (+10)$

   $= (-14) + (-9) + (+10)$

   $= (-23) + (+10)$

   $= -13$

11. $(+5) - (-3) = (+5) + (+3) = 8$

13. $(-30) - (+3) = (-30) + (-3) = -33$

15. $\left(-\dfrac{7}{8}\right) + \left(-\dfrac{3}{4}\right) = \left(-\dfrac{7}{8}\right) + \left(-\dfrac{6}{8}\right) = -\dfrac{13}{8} = -1\dfrac{5}{8}$

17. $-20.8 - 1.9 = -20.8 + (-1.9) = -22.7$

19. $-5 + (-2) - (-3) = -5 + (-2) + (+3)$

    $= (-7) + (+3) = -4$

21. $(-16) + (-13) = -29$

23. $(-7) \times 6 = -42$

25. $(-3)(-2) + (-2)(4) = 6 + (-8) = -2$

27. $(-20) \div 4 = \dfrac{-20}{4} = -5$

29. $\left(-\dfrac{1}{2}\right) \div \dfrac{3}{4} = \left(-\dfrac{1}{2}\right)\left(\dfrac{2 \cdot 2}{3}\right) = -\dfrac{2}{3}$

31. $(-5)(-1)(2) = (5)(1)(2) = 10$

33. $(-1)(-2)(-3)(-5) = (1)(2)(3)(5) = 30$

35. $-\dfrac{4}{3} + \dfrac{2}{3} + \dfrac{1}{6} = -\dfrac{2}{3} + \dfrac{1}{6} = -\dfrac{4}{6} + \dfrac{1}{6} = -\dfrac{3}{6} = -\dfrac{1}{2}$

37. $(-3)(-2)(-5) = -(3)(2)(5) = -30$

39. $3.5(-2.6) = -9.1$

41. $5 - (-3.5) + 1.6 = 5 + 3.5 + 1.6 = 10.1$

43. $25 + 4.1 + (-26) + (-3.5)$

    $= 29.1 + (-29.5) = -0.4$

45. $(-8)(3) = -24$; 24 yards lost

47. $6895 - (-468) = 6895 + 468 = 7363$ ft.

49. $(-3)^4 = (-3)(-3)(-3)(-3) = 81$

51. $(-5)^3 = (-5)(-5)(-5) = -125$

53. $-8^2 = -8 \cdot 8 = -64$

55. $\left(\dfrac{5}{6}\right)^2 = \left(\dfrac{5}{6}\right)\left(\dfrac{5}{6}\right) = \dfrac{25}{36}$

57. $5(3x - 7y) = 5(3x) + 5(-7y) = 15x - 35y$

59. $-(7x^2 - 3x + 11)$

    $= -1(7x^2) + (-1)(-3x) + (-1)(11)$

    $= -7x^2 + 3x - 11$

61. $3a^2b - 2bc + 6bc^2 - 8a^2b - 6bc^2 + 5bc$

    $= (3 - 8)a^2b + (-2 + 5)bc + (6 - 6)bc^2$

    $= -5a^2b + 3bc$

63. $4x^2 - 13x + 7 - 9x^2 - 22x - 16$

$\quad = (4-9)x^2 + (-13-22)x + 7 - 16$

$\quad = -5x^2 - 35x - 9$

65. $(5)(-4) + (3)(-2)^3$

$\quad = (5)(-4) + (3)(-8)$

$\quad = -20 + (-24)$

$\quad = -44$

67. $(7-9)^3 + (-6)(-2) + (-3)$

$\quad = (-2)^3 + (-6)(-2) + (-3)$

$\quad = -8 + (-6)(-2) + (-3)$

$\quad = -8 + (12) + (-3)$

$\quad = 1$

69. If $x = 6$, then $5 - \dfrac{2}{3}x$

$\quad = 5 - \dfrac{2}{3}(6) = 5 - 4 = 1$

71. If $x = 2$, then $-3x^2 - 4x + 5$

$\quad = -3(2)^2 - 4(2) + 5$

$\quad = -12 - 8 + 5 = -15$

73. If $v = 24$, $t = 2$, and $a = 32$, then

$\quad d = vt - \dfrac{1}{2}at^2 = 24(2) - \dfrac{1}{2}(32)(2)^2 = -16$

75. If $p = 6000$, $r = 18\%$, and $t = \dfrac{3}{4}$, then

$\quad I = prt = 6000(0.18)\left(\dfrac{3}{4}\right) = \$810$

77. If $r = 15$, $A = \pi r^2 = 3.14(15)^2$

$\qquad\qquad\qquad = 706.5$ sq. m.

$\quad \text{Cost} = (706.5 \text{ sq. m})(\$3 / \text{sq. m})$

$\qquad\quad = \$2119.50$

79. $3x - 2(x + 4) = 3x - 2x - 8 = x - 8$

81. $2[3 - (4 - 5x)]$

$\quad = 2(3 - 4 + 5x)$

$\quad = 2(-1 + 5x)$

$\quad = -2 + 10x$

83. $2xy^3 - 6x^3y - 4x^2y^2 + 3(xy^3 - 2x^2y - 3x^2y^2)$

$\quad = 2xy^3 - 6x^3y - 4x^2y^2 + 3xy^3 - 6x^2y - 9x^2y^2$

$\quad = (2+3)xy^3 - 6x^3y + (-4-9)x^2y^2 - 6x^2y$

$\quad = 5xy^3 - 6x^3y - 13x^2y^2 - 6x^2y$

85. $-5(x + 2y - 7) + 3x(2 - 5y)$

$\quad = -5x - 10y + 35 + 6x - 15xy$

$\quad = x - 10y + 35 - 15xy$

87. $8[a - b(3 - 4a)] - 6a[2 - a(3 - b)]$

$\quad = 8(a - 3b + 4ab) - 6a(2 - 3a + ab)$

$\quad = 8a - 24b + 32ab - 12a + 18a^2 - 6a^2b$

$\quad = -4a - 24b + 32ab + 18a^2 - 6a^2b$

89. $-3\{2x - [x - 3y(x - 2y)]\}$

$\quad = -3[2x - (x - 3xy + 6y^2)]$

$\quad = -3(2x - x + 3xy - 6y^2)$

$\quad = -3(x + 3xy - 6y^2)$

$\quad = -3x - 9xy + 18y^2$

Test - Chapter 1

1. $-3 + (-4) + 9 + 2 = -7 + 11 = 4$

3. $-0.6 - (-0.8) = -0.6 + (0.8) = 0.2$

5. $(-5)(-2)(7)(-1) = -(5)(2)(7)(1) = -70$

7. $(-1.8) \div (0.6) = \dfrac{-1.8}{0.6} = -3$

9. $(-4)^3 = (-4)(-4)(-4) = -64$

11. $-5^3 = -(5)(5)(5) = -125$

13. $-5x(x + 2y - 7) = -5x^2 - 10xy + 35x$

15. $2xy - 5x^2y + 3xy^2 - 8xy - 7xy^2$

    $= (2-8)xy - 5x^2y + (3-7)xy^2$

    $= -6xy - 5x^2y - 4xy^2$

17. $12a(a+b) - 4(a^2 - 2ab)$

    $= 12a^2 + 12ab - 4a^2 + 8ab$

    $= (12-4)a^2 + (12+8)ab$

    $= 8a^2 + 20ab$

19. $5(3x - 2y) - (x + 6y)$

    $= 15x - 10y - x - 6y$

    $= (15-1)x + (-10-6)y$

    $= 14x - 16y$

21. $3(4-6)^3 + 12 \div (-4) + 2$

    $= 3(-2)^3 + 12 \div (-4) + 2$

    $= 3(-8) + 12 \div (-4) + 2$

    $= -24 - 3 + 2$

    $= -25$

23. If $x = -3$, then $3x^2 - 7x - 11$

    $= 3(-3)^2 - 7(-3) - 11$

    $= 37$

25. (60 miles / hr)(1.61 km / mile)

    $= 96.6$ km / hr

27. $3\{[x - (5 - 2y)] - 4[3 + (6x - 7y)]\}$

    $= 3[(x - 5 + 2y) - 4(3 + 6x - 7y)]$

    $= 3(x - 5 + 2y - 12 - 24x + 28y)$

    $= 3(-23x + 30y - 17)$

    $= -69x + 90y - 51$

Practice Quiz: Sections 1.1 - 1.5

1. $(-5) - (-4)$

3. $(-3)(5)$

5. $\left(-\dfrac{2}{5}\right)^3$

Practice Quiz: Sections 1.6 - 1.9

Combine like terms.

  1. $2x^2 - 3y^2 - 5x^2 + y^2$

Simplify.

  3. $4(7-2)^2 + 12 \div 2 - 3$

  5. Simplify $5x - 3[2 - 5(x - 4)]$

Answers to Practice Quiz

Sections 1.1 - 1.5

  1. -1

  3. -15

  5. $-\dfrac{8}{125}$

Answers to Practice Quiz

Sections 1.6 - 1.9

  1. $-3x^2 - 2y^2$

  3. 103

  5. $20x - 66$

Pretest Chapter 2

1. $$x - 9 = 49$$
$$x - 9 + 9 = 49 + 9$$
$$x = 58$$

3. $$-5x = 20$$
$$\frac{-5x}{-5} = \frac{20}{-5}$$
$$x = -4$$

5. $$3x - 8 = 7x + 6$$
$$3x + (-3x) - 8 = 7x + (-3x) + 6$$
$$-8 = 4x + 6$$
$$-8 + (-6) = 4x + 6 + (-6)$$
$$-14 = 4x$$
$$\frac{-14}{4} = \frac{4x}{4}$$
$$-\frac{7}{2} = x$$

7. $$\frac{2}{5}x + \frac{1}{4} = \frac{1}{2}x$$
$$20\left(\frac{2}{5}x\right) + 20\left(\frac{1}{4}\right) = 20\left(\frac{1}{2}x\right)$$
$$8x + 5 = 10x$$
$$8x + (-8x) + 5 = 10x + (-8x)$$
$$5 = 2x$$
$$\frac{5}{2} = x$$

9. (a) $C = \frac{5}{9}(F - 32)$
$$\frac{9}{5}C = \frac{9}{5}\left(\frac{5}{9}\right)(F - 32)$$
$$\frac{9}{5}C = F - 32$$
$$\frac{9}{5}C + 32 = F - 32 + 32$$
$$\frac{9}{5}C + 32 = F$$

   (b) $C = -15°$
$$\frac{9}{5}(-15) + 32 = F$$
$$-27 + 32 = F$$
$$5° = F$$

11. $1 < 10$

13. $-12 > -13$

15. $$-2x + 5 \leq 4 - x + 3$$
$$-2x + 5 \leq -x + 7$$
$$-2x + x + 5 \leq -x + x + 7$$
$$-x + 5 \leq 7$$
$$-x + 5 + (-5) \leq 7 + (-5)$$
$$-x \leq 2$$
$$x \geq -2$$

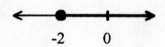

Exercises 2.1

1. When we use the <u>equal</u> sign, we indicate two expressions are <u>equal</u> in value.

3. The <u>solution</u> of an equation is a value of the variable that makes the equation true.

5. Answers may vary. A sample answer is "to isolate the variable".

7. $$x + 3 = 6$$
$$x + 3 + (-3) = 6 + (-3)$$
$$x = 3$$
Check: $3 + 3 \overset{?}{=} 6$
$$6 = 6$$

9. $$6 = x + 7$$
$$6 + (-7) = x + 7 + (-7)$$
$$-1 = x$$
Check: $6 \overset{?}{=} -1 + 7$
$$6 = 6$$

11. $$x - 18 = 2$$
$$x - 18 + 18 = 2 + 18$$
$$x = 20$$
Check: $20 - 18 \overset{?}{=} 2$
$$2 = 2$$

24

13. $$0 = x - 7$$
$$0 + 7 = x - 7 + 7$$
$$7 = x$$

Check: $0 \overset{?}{=} 7 - 7$
$$0 = 0$$

15. $$8 - 2 = x + 5$$
$$6 = x + 5$$
$$6 + (-5) = x + 5 + (-5)$$
$$1 = x$$

Check: $8 - 2 \overset{?}{=} 1 + 5$
$$6 = 6$$

17. $$18 - 2 + 3 = x + 19$$
$$19 = x + 19$$
$$19 + (-19) = x + 19 + (-19)$$
$$0 = x$$

Check $18 - 2 + 3 \overset{?}{=} 0 + 19$
$$19 = 19$$

19. $$x - 5 = -26$$
$$x - 5 + 5 = -26 + 5$$
$$x = -21$$

Check: $-21 - 5 \overset{?}{=} -26$
$$-26 = -26$$

21. $$-18 + x = 48$$
$$-18 + 18 + x = 48 + 18$$
$$x = 66$$

Check: $-18 + 66 \overset{?}{=} 48$
$$48 = 48$$

23. $$23 - 8 = x - 12$$
$$15 = x - 12$$
$$15 + 12 = x - 12 + 12$$
$$27 = x$$

Check: $23 - 8 \overset{?}{=} 27 - 12$
$$15 = 15$$

25. $$x + 7 = 15 - 26 + 4$$
$$x + 7 = -7$$
$$x + 7 + (-7) = -7 + (-7)$$
$$x = -14$$

Check: $-14 + 7 \overset{?}{=} 15 - 26 + 4$
$$-7 = -7$$

27. $$-19 + x = 7, \quad x \overset{?}{=} 12$$
$$-19 + 12 \overset{?}{=} 7$$
$$-7 = 7$$
$$x = 12 \text{ is not the solution.}$$
$$-19 + x = 7$$
$$-19 + 19 + x = 7 + 19$$
$$x = 26$$

29. $$-10 - 3 = x - 20, \quad x \overset{?}{=} -3$$
$$-10 - 3 \overset{?}{=} -3 - 20$$
$$-13 = -23$$
$$x = -3 \text{ is not the solution.}$$
$$-10 - 3 = x - 20$$
$$-13 = x - 20$$
$$-13 + 20 = x - 20 + 20$$
$$7 = x$$

31. $$-39 = x - 47, \quad x \overset{?}{=} -8$$
$$-39 \overset{?}{=} -8 - 47$$
$$-39 = -55$$
$$x = -8 \text{ is not the solution.}$$
$$-39 = x - 47$$
$$-39 + 47 = x - 47 + 47$$
$$8 = x$$

33. $$x + 8 = 12 - 19 + 3, \quad x \overset{?}{=} -12$$
$$-12 + 8 \overset{?}{=} 12 - 19 + 3$$
$$-4 = -4$$
$$x = -4 \text{ is the solution.}$$

35. $$-9 + x = -4$$
$$-9 + 9 + x = -4 + 9$$
$$x = 5$$

37. $$27 = 5 + x$$
$$27 + (-5) = 5 + (-5) + x$$
$$22 = x$$

39. $$3.6 + 1.2 = x + 1.3$$
$$4.8 = x + 1.3$$
$$4.8 + (-1.3) = x + 1.3 + (-1.3)$$
$$3.5 = x$$

41. $$-0.6 + x = -1.8$$
$$-0.6 + 0.6 + x = -1.8 + 0.6$$
$$x = -1.2$$

43. $$x + \frac{1}{3} = \frac{2}{3}$$
$$x + \frac{1}{3} + \left(-\frac{1}{3}\right) = \frac{2}{3} + \left(-\frac{1}{3}\right)$$
$$x = \frac{1}{3}$$

45. $$x - \frac{1}{5} = \frac{1}{2} + \frac{1}{10}$$
$$x - \frac{2}{10} = \frac{5}{10} + \frac{1}{10}$$
$$x - \frac{2}{10} = \frac{6}{10}$$
$$x - \frac{2}{10} + \frac{2}{10} = \frac{6}{10} + \frac{2}{10}$$
$$x = \frac{8}{10}$$
$$x = \frac{4}{5}$$

47. $$\frac{3}{22} - \frac{7}{11} = x - \frac{3}{2}$$
$$\frac{3}{22} - \frac{14}{22} = x - \frac{33}{22}$$
$$-\frac{11}{22} = x - \frac{33}{22}$$
$$\frac{-11}{22} + \frac{33}{22} = x - \frac{33}{22} + \frac{33}{22}$$
$$\frac{22}{22} = x$$
$$1 = x$$

49. $$7\frac{1}{8} = -20 + x$$
$$7\frac{1}{8} + 20 = -20 + 20 + x$$
$$27\frac{1}{8} = x$$

51. $$1.6 - 5x + 3.2 = -2x + 5.6 + 4x - 8x$$
$$-5x + 4.8 = -6x + 5.6$$
$$-5x + 6x + 4.8 = -6x + 6x + 5.6$$
$$x + 4.8 = 5.6$$
$$x + 4.8 + (-4.8) = 5.6 + (-4.8)$$
$$x = 0.8$$

53. $$x + 9.3715 = -18.1261$$
$$x + 9.3715 + (-9.3715) = -18.1261 + (-9.3715)$$
$$x = -27.4976$$

Cumulative Review Problems

55. $$x + 3y - 5x - 7y + 2x$$
$$= (1 - 5 + 2)x + (3 - 7)y$$
$$= -2x - 4y$$

57. $$y^2 + y - 12 - 3y^2 - 5y + 16$$
$$= (1 - 3)y^2 + (1 - 5)y - 12 + 16$$
$$= 2y^2 - 4y + 4$$

Exercises 2.2

1. Divide each side of
$-5x = -30$ by $\underline{-5}$.

3. Multiply each side of
$\frac{1}{7}x = -2$ by $\underline{7}$.

5. $$\frac{1}{3}x = 21$$
$$3\left(\frac{1}{3}x\right) = 3(21)$$
$$x = 63$$

Check: $\frac{1}{3}(63) \overset{?}{=} 21$
$$21 = 21$$

26

7. $\dfrac{1}{4}x = -2$

$4\left(\dfrac{1}{4}x\right) = 4(-2)$

$x = -8$

Check: $\dfrac{1}{4}(-8)\overset{?}{=} -2$

$-2 = -2$

9. $\dfrac{x}{5} = 16$

$5\left(\dfrac{x}{5}\right) = 5(16)$

$x = 80$

Check: $\dfrac{80}{5}\overset{?}{=} 16$

$16 = 16$

11. $-4 = \dfrac{x}{12}$

$12(-4) = 12\left(\dfrac{x}{12}\right)$

$-48 = x$

Check: $-4 \overset{?}{=} \dfrac{-48}{12}$

$-4 = -4$

13. $7x = 28$

$\dfrac{7x}{7} = \dfrac{28}{7}$

$x = 4$

Check: $7(4)\overset{?}{=} 28$

$28 = 28$

15. $-16 = 6x$

$\dfrac{-16}{6} = \dfrac{6x}{6}$

$-\dfrac{8}{3} = x$

17. $1.5x = 75$

$\dfrac{1.5x}{1.5} = \dfrac{75}{1.5}$

$x = 50$

19. $-15 = -x$

$\dfrac{-15}{-1} = \dfrac{-x}{-1}$

$15 = x$

21. $-84 = 12x$

$\dfrac{-84}{12} = \dfrac{12x}{12}$

$-7 = x$

23. $0.5x = 0.20$

$\dfrac{0.5x}{0.5} = \dfrac{0.20}{0.5}$

$x = 0.4$

25. $-3x = 21, \quad x \overset{?}{=} 7$

$-3(7)\overset{?}{=} 21$

$-21 \neq 21$

$x = 7$ is not the solution.

$-3x = 21$

$\dfrac{-3x}{-3} = \dfrac{21}{-3}$

$x = -7$

27. $-11x = 66, \quad x \overset{?}{=} -6$

$-11(-6)\overset{?}{=} 66$

$66 = 66$

$x = -6$ is the solution.

29. $-3y = 2.4$

$\dfrac{-3y}{-3} = \dfrac{2.4}{-3}$

$y = -0.8$

31. $-27 = -12z$

$\dfrac{-27}{-12} = \dfrac{-12z}{-12}$

$\dfrac{9}{4} = z$

33. $-5.2y = 3.12$

$\dfrac{-5.2y}{-5.2} = \dfrac{3.12}{-5.2}$

$y = -0.6$

35. $4x + 3x = 21$

$\qquad 7x = 21$

$\qquad \dfrac{7x}{7} = \dfrac{21}{7}$

$\qquad x = 3$

37. $2x - 7x = 20$

$\qquad -5x = 20$

$\qquad \dfrac{-5x}{-5} = \dfrac{20}{-5}$

$\qquad x = -4$

39. $-12y + y = -15$

$\qquad -11y = -15$

$\qquad \dfrac{-11y}{-11} = \dfrac{-15}{-11}$

$\qquad y = \dfrac{15}{11}$

41. $\qquad -\dfrac{2}{3} = -\dfrac{4}{7}x$

$\quad -\dfrac{7}{4}\left(-\dfrac{2}{3}\right) = -\dfrac{7}{4}\left(-\dfrac{4}{7}x\right)$

$\qquad \dfrac{7}{6} = x$

43. $3.6172x = -19.026472$

$\quad \dfrac{3.6172x}{3.6172} = \dfrac{-19.026472}{3.6172}$

$\qquad x = -5.26$

Cumulative Review Problems

45. $5 - 3(2) = 5 - 6 = -1$

47. $(-3)^3 + (-20) \div 2$

$= -27 + (-20) \div 2$

$= -27 + (-10)$

$= -37$

1. $\qquad 7x + 9 = 51$

$\quad 7x + 9 + (-9) = 51 + (-9)$

$\qquad 7x = 42$

$\qquad \dfrac{7x}{7} = \dfrac{42}{7}$

$\qquad x = 6$

Check: $7(6) + 9 \overset{?}{=} 51$

$\qquad 42 + 9 \overset{?}{=} 51$

$\qquad 51 = 51$

3. $\qquad 6x - 4 = 62$

$\quad 6x - 4 + 4 = 62 + 4$

$\qquad 6x = 66$

$\qquad \dfrac{6x}{6} = \dfrac{66}{6}$

$\qquad x = 11$

Check: $6(11) - 4 \overset{?}{=} 62$

$\qquad 66 - 4 \overset{?}{=} 62$

$\qquad 62 = 62$

5. $\qquad 4x - 13 = -81$

$\quad 4x - 13 + 13 = -81 + 13$

$\qquad 4x = -68$

$\qquad \dfrac{4x}{4} = -\dfrac{68}{4}$

$\qquad x = -17$

Check: $4(-17) - 13 \overset{?}{=} -81$

$\qquad -68 - 13 \overset{?}{=} -81$

$\qquad -81 = -81$

7.
$$-5x + 14 = -41$$
$$-5x + 14 + (-14) = -41 + (-14)$$
$$-5x = -55$$
$$\frac{-5x}{-5} = \frac{-55}{-5}$$
$$x = 11$$

Check: $-5(11) + 14 \overset{?}{=} -41$

$$-55 + 14 \overset{?}{=} -41$$
$$-41 = -41$$

9.
$$-15 + 2x = 15$$
$$-15 + 15 + 2x = 15 + 15$$
$$2x = 30$$
$$\frac{2x}{2} = \frac{30}{2}$$
$$x = 15$$

Check: $-15 + 2(15) \overset{?}{=} 15$

$$-15 + 30 \overset{?}{=} 15$$
$$15 = 15$$

11.
$$\frac{1}{2}x - 3 = 11$$
$$\frac{1}{2}x - 3 + 3 = 11 + 3$$
$$\frac{1}{2}x = 14$$
$$2\left(\frac{1}{2}x\right) = 2(14)$$
$$x = 28$$

Check: $\frac{1}{2}(28) - 3 \overset{?}{=} 11$

$$14 - 3 \overset{?}{=} 11$$
$$11 = 11$$

13.
$$\frac{1}{6}x + 2 = -4$$
$$\frac{1}{6}x + 2 + (-2) = -4 + (-2)$$
$$\frac{1}{6}x = -6$$
$$6\left(\frac{1}{6}x\right) = 6(-6)$$
$$x = -36$$

Check: $\frac{1}{6}(-36) + 2 \overset{?}{=} -4$

$$-6 + 2 \overset{?}{=} -4$$
$$-4 = -4$$

15.
$$8x = 48 + 2x$$
$$8x + (-2x) = 48 + 2x + (-2x)$$
$$6x = 48$$
$$\frac{6x}{6} = \frac{48}{6}$$
$$x = 8$$

Check: $8(8) \overset{?}{=} 48 + 2(8)$

$$64 \overset{?}{=} 48 + 16$$
$$64 = 64$$

17.
$$-6x = -27 + 3x$$
$$-6x + (-3x) = -27 + 3x + (-3x)$$
$$-9x = -27$$
$$\frac{-9x}{-9} = \frac{-27}{-9}$$
$$x = 3$$

Check: $-6(3) \overset{?}{=} -27 + 3(3)$

$$-18 \overset{?}{=} -27 + 9$$
$$-18 = -18$$

19. $30 - x = 5x$

$30 - x + x = 5x + x$

$30 = 6x$

$\dfrac{30}{6} = \dfrac{6x}{6}$

$5 = x$

Check: $30 - 5 \overset{?}{=} 5(5)$

$25 = 25$

21. $54 - 2x = -8x$

$54 - 2x + 2x = -8x + 2x$

$54 = -6x$

$\dfrac{54}{-6} = \dfrac{-6x}{-6}$

$-9 = x$

Check: $54 - 2(-9) \overset{?}{=} -8(-9)$

$54 + 18 \overset{?}{=} 72$

$72 = 72$

23. $5y + 2 = 6y - 6 + y, \quad y \overset{?}{=} 4$

$5(4) + 2 \overset{?}{=} 6(4) - 6 + 4$

$20 + 2 \overset{?}{=} 24 - 2$

$22 = 22$

$y = 4$ is the solution.

25. $9x + 2 - 5x = -8 + 5x - 2, \quad x \overset{?}{=} -12$

$9(-12) + 2 - 5(-12) \overset{?}{=} -8 + 5(-12) - 2$

$-108 + 2 + 60 \overset{?}{=} -8 - 60 - 2$

$-46 \neq -70$

$x = -12$ is not the solution.

$9x + 2 - 5x = -8 + 5x - 2$

$4x + 2 = 5x - 10$

$4x + (-4x) + 2 = 5x + (-4x) - 10$

$2 = x - 10$

$2 + 10 = x - 10 + 10$

$12 = x$

27. $7y + 21 - 5y = 5y - 7 + y$

Left: $2y + 21 = 6y - 7$

$2y + (-6y) + 21 = 6y + (-6y) - 7$

$-4y + 21 = -7$

$-4y + 21 + (-21) = -7 + (-21)$

$-4y = -28$

$\dfrac{-4y}{-4} = \dfrac{-28}{-4}$

$y = 7$

Right: $2y + 21 = 6y - 7$

$2y + (-2y) + 21 = 6y + (-2y) - 7$

$21 = 4y - 7$

$21 + 7 = 4y - 7 + 7$

$28 = 4y$

$\dfrac{28}{4} = \dfrac{4y}{4}$

$7 = y$

Neither approach is better.

29. $5x + 6 = -7x - 4$

$5x + 7x + 6 = -7x + 7x - 4$

$12x + 6 = -4$

$12x + 6 + (-6) = -4 + (-6)$

$12x = -10$

$\dfrac{12x}{12} = \dfrac{-10}{12}$

$x = -\dfrac{5}{6}$

31. $2x + 5 = 4x - 5$

$2x + (-4x) + 5 = 4x + (-4x) - 5$

$-2x + 5 = -5$

$-2x + 5 + (-5) = -5 + (-5)$

$-2x = -10$

$\dfrac{-2x}{-2} = \dfrac{-10}{-2}$

$x = 5$

33.
$$5y - 7 = 3y - 9$$
$$5y + (-3y) - 7 = 3y + (-3y) - 9$$
$$2y - 7 = -9$$
$$2y - 7 + 7 = -9 + 7$$
$$2y = -2$$
$$\frac{2y}{2} = \frac{-2}{2}$$
$$y = -1$$

35.
$$9x - 5 + 4x = 7x + 43 - 2x$$
$$13x - 5 = 5x + 43$$
$$13x + (-5x) - 5 = 5x + (-5x) + 43$$
$$8x - 5 = 43$$
$$8x - 5 + 5 = 43 + 5$$
$$8x = 48$$
$$\frac{8x}{8} = \frac{48}{8}$$
$$x = 6$$

37.
$$3(2y - 4) = 12$$
$$6y - 12 = 12$$
$$6y = 24$$
$$y = 4$$
Check: $3[2(4) - 4] \overset{?}{=} 12$
$$3(8 - 4) \overset{?}{=} 12$$
$$12 = 12$$

39.
$$4(2x + 1) - 7 = 6 - 5$$
$$8x + 4 - 7 = 6 - 5$$
$$8x - 3 = 1$$
$$8x = 4$$
$$x = \frac{1}{2}$$
Check: $4\left[2\left(\frac{1}{2}\right) + 1\right] - 7 \overset{?}{=} 6 - 5$
$$4(1 + 1) - 7 \overset{?}{=} 1$$
$$1 = 1$$

41.
$$6(3 - 4x) + 17 = 8x - 3(2 - 3x)$$
$$18 - 24x + 17 = 8x - 6 + 9x$$
$$-24x + 35 = 17x - 6$$
$$41 = 41x$$
$$1 = x$$
Check: $6[3 - 4(1)] + 17 \overset{?}{=} 8(1) - 3[2 - 3(1)]$
$$6(3 - 4) + 17 \overset{?}{=} 8 - 3(2 - 3)$$
$$-6 + 17 \overset{?}{=} 8 + 3$$
$$11 = 11$$

43.
$$3(x + 0.2) = 2(x - 0.3) + 5.2$$
$$3x + 0.6 = 2x - 0.6 + 5.2$$
$$x + 0.6 = 4.6$$
$$x = 4$$
Check: $3(4 + 0.2) \overset{?}{=} 2(4 - 0.3) + 5.2$
$$3(4.2) \overset{?}{=} 2(3.7) + 5.2$$
$$12.6 = 12.6$$

45.
$$5x - (2x - 3) = 4(x + 9)$$
$$5x - 2x + 3 = 4x + 36$$
$$3x + 3 = 4x + 36$$
$$-33 = x$$
Check: $5(-33) - [2(-33) - 3] \overset{?}{=} 4[(-33) + 9]$
$$-165 - (-69) \overset{?}{=} 4(-24)$$
$$-96 = -96$$

47.
$$-3(y - 3y) + 4 = -4(3y - y) + 6 + 13y$$
$$-3(-2y) + 4 = -4(2y) + 6 + 13y$$
$$6y + 4 = 5y + 6$$
$$y = 2$$
Check: $-3[2 - 3(2)] + 4 \overset{?}{=} -4[3(2) - 2] + 6 + 13(2)$
$$-3(-4) + 4 \overset{?}{=} -4(4) + 6 + 26$$
$$16 = 16$$

49. $0.2(x+3)-(x-1.5)=0.3(x+2)-2.9$

$0.2x+0.6-x+1.5=0.3x+0.6-2.9$

$-0.8x+2.1=0.3x-2.3$

$4.4=1.1x$

$4=x$

Check: $0.2(4+3)-(4-1.5)\overset{?}{=}0.3(4+2)-2.9$

$1.4-2.5\overset{?}{=}1.8-2.9$

$-1.1=-1.1$

51. $1.2x+4=3.2x-8$

$12=2x$

$6=x$

53. $5z+7-2z=32-2z$

$3z+7=32-2z$

$5z=25$

$z=5$

55. $-4w-28=-7-w$

$-21=3w$

$-7=w$

57. $9x-16+6x=11+4x-5$

$15x-16=4x+6$

$11x=22$

$x=2$

59. $2x^2-3x-8=2x^2+5x-6$

$-2=8x$

$-\dfrac{1}{4}=x$

61. $-3.5x+1.3=-2.7x+1.5$

$-0.2=0.8x$

$-0.25=x$

63. $5(4+x)=3(3x-1)-9$

$20+5x=9x-3-9$

$20+5x=9x-12$

$32=4x$

$8=x$

65. $17(y+3)-4(y-10)=13$

$17y+51-4y+40=13$

$13y+91=13$

$13y=-78$

$y=-6$

67. $10(x-5)-2(x-9)=-28$

$10x-50-2x+18=-28$

$8x-32=-28$

$8x=4$

$x=\dfrac{1}{2}$

69. $1.63x-9.23=5.71x+8.04$

$-17.27=4.08x$

$-4.23=x$

Cumulative Review Problems

71. $2x(3x-y)+4(2x^2-3xy)$

$=6x^2-2xy+8x^2-12xy$

$=14x^2-14xy$

73. $2\{3-2[x-4(2x+3)]\}$

$=2[3-2(x-8x-12)]$

$=2[3-2(-7x-12)]$

$=2(3+14x+24)$

$=2(14x+27)$

$=28x+54$

Exercises 2.4

1. $\dfrac{1}{5}x+\dfrac{1}{10}=\dfrac{1}{2}$

$10\left(\dfrac{1}{5}x\right)+10\left(\dfrac{1}{10}\right)=10\left(\dfrac{1}{2}\right)$

$2x+1=5$

$2x=4$

$x=2$

Check: $\dfrac{1}{5}(2)+\dfrac{1}{10}\overset{?}{=}\dfrac{1}{2}$

$\dfrac{4}{10}+\dfrac{1}{10}\overset{?}{=}\dfrac{1}{2}$

$\dfrac{1}{2}=\dfrac{1}{2}$

3. $\dfrac{2}{3}x = \dfrac{1}{3}x + \dfrac{1}{6}$

$6\left(\dfrac{2}{3}x\right) = 6\left(\dfrac{1}{3}x\right) + 6\left(\dfrac{1}{6}\right)$

$4x = 2x + 1$

$2x = 1$

$x = \dfrac{1}{2}$

Check: $\dfrac{2}{3}\left(\dfrac{1}{2}\right) \overset{?}{=} \dfrac{1}{3}\left(\dfrac{1}{2}\right) + \dfrac{1}{6}$

$\dfrac{1}{3} \overset{?}{=} \dfrac{1}{6} + \dfrac{1}{6}$

$\dfrac{1}{3} = \dfrac{1}{3}$

5. $\dfrac{y}{2} + \dfrac{y}{3} = \dfrac{5}{6}$

$6\left(\dfrac{y}{2}\right) + 6\left(\dfrac{y}{3}\right) = 6\left(\dfrac{5}{6}\right)$

$3y + 2y = 5$

$5y = 5$

$y = 1$

Check: $\dfrac{1}{2} + \dfrac{1}{3} \overset{?}{=} \dfrac{5}{6}$

$\dfrac{3}{6} + \dfrac{2}{6} \overset{?}{=} \dfrac{5}{6}$

$\dfrac{5}{6} = \dfrac{5}{6}$

7. $20 - \dfrac{1}{3}x = \dfrac{1}{2}x$

$6(20) - 6\left(\dfrac{1}{3}x\right) = 6\left(\dfrac{1}{2}x\right)$

$120 - 2x = 3x$

$120 = 5x$

$24 = x$

Check: $20 - \dfrac{1}{3}(24) \overset{?}{=} \dfrac{1}{2}(24)$

$20 - 8 \overset{?}{=} 12$

$12 = 12$

9. $2 + \dfrac{y}{2} = \dfrac{3y}{4} - 3$

$4(2) + 4\left(\dfrac{y}{2}\right) = 4\left(\dfrac{3y}{4}\right) - 4(3)$

$8 + 2y = 3y - 12$

$20 = y$

Check: $2 + \left(\dfrac{20}{2}\right) \overset{?}{=} \dfrac{3(20)}{4} - 3$

$2 + 10 \overset{?}{=} 15 - 3$

$12 = 12$

11. $\dfrac{y-1}{2} = 4 - \dfrac{y}{7}$

$14\left(\dfrac{y-1}{2}\right) = 14(4) - 14\left(\dfrac{y}{7}\right)$

$7y - 7 = 56 - 2y$

$9y = 63$

$y = 7$

Check: $\dfrac{7-1}{2} \overset{?}{=} 4 - \dfrac{7}{7}$

$\dfrac{6}{2} \overset{?}{=} 4 - 1$

$3 = 3$

13. $0.5 - 0.3x = 3.8$

$10(0.5) - 10(0.3x) = 10(3.8)$

$5 - 3x = 38$

$-3x = 33$

$x = -11$

Check: $0.5 - 0.3(-11) \overset{?}{=} 3.8$

$0.5 + 3.3 \overset{?}{=} 3.8$

$3.8 = 3.8$

15. $5.9 = 2.5x - 1.6$

$10(5.9) = 10(2.5x) - 10(1.6)$

$59 = 25x - 16$

$75 = 25x$

$3 = x$

Check: $5.9 \overset{?}{=} 2.5(3) - 1.6$

$5.9 \overset{?}{=} 7.5 - 1.6$

$5.9 = 5.9$

17. $\frac{1}{2}(y-2)+2=\frac{3}{8}(3y-4),\ y\overset{?}{=}4$

$\frac{1}{2}(4-2)+2\overset{?}{=}\frac{3}{8}(3\cdot4-4)$

$1+2\overset{?}{=}\frac{3}{8}(8)$

$3=3$

$y=4$ is a solution.

19. $\frac{1}{2}\left(y-\frac{1}{5}\right)=\frac{1}{5}(y+2),\ y\overset{?}{=}\frac{5}{8}$

$\frac{1}{2}\left(\frac{5}{8}-\frac{1}{5}\right)\overset{?}{=}\frac{1}{5}\left(\frac{5}{8}+2\right)$

$\frac{1}{2}\left(\frac{17}{40}\right)\overset{?}{=}\frac{1}{5}\left(\frac{21}{8}\right)$

$\frac{17}{80}\neq\frac{42}{80}$

$y=\frac{5}{8}$ is not a solution.

21. $4(3x-2)=\frac{1}{3}(x-1)+4$

$12x-8=\frac{1}{3}x-\frac{1}{3}+4$

$3(12x)-3(8)=3\left(\frac{1}{3}x\right)-3\left(\frac{1}{3}\right)+3(4)$

$36x-24=x-1+12$

$36x-24=x+11$

$35x=35$

$x=1$

23. $0.7-0.3x-0.6=0.1$

$10(0.7)-10(0.3x)-10(0.6)=10(0.1)$

$7-3x-6=1$

$-3x+1=1$

$-3x=0$

$x=0$

25. $-0.3(x-1)+2.0=-0.4(2x+3)$

$10(-0.3)(x-1)+10(2.0)=10(-0.4)(2x+3)$

$-3(x-1)+20=-4(2x+3)$

$-3x+3+20=-8x-12$

$-3x+23=-8x-12$

$5x=-35$

$x=-7$

27. $-5(0.2x+0.1)-0.6=1.9$

$-x-0.5-0.6=1.9$

$-x-1.1=1.9$

$-x=3$

$x=-3$

29. $\frac{1}{3}(y+2)=3y-5(y-2)$

$3\left(\frac{1}{3}\right)(y+2)=3(3y)-3(5)(y-2)$

$y+2=9y-15y+30$

$y+2=-6y+30$

$7y=28$

$y=4$

31. $\frac{1+2x}{5}+\frac{4-x}{3}=\frac{1}{15}$

$15\left(\frac{1+2x}{5}\right)+15\left(\frac{4-x}{3}\right)=15\left(\frac{1}{15}\right)$

$3+6x+20-5x=1$

$x+23=1$

$x=-22$

33. $\frac{x}{5}-\frac{2}{3}x+\frac{16}{15}=\frac{1}{3}(x-4)$

$15\left(\frac{x}{5}\right)-15\left(\frac{2}{3}x\right)+15\left(\frac{16}{15}\right)=15\left(\frac{1}{3}\right)(x-4)$

$3x-10x+16=5x-20$

$-7x+16=5x-20$

$-12x=-36$

$x=3$

35. $\dfrac{1}{3}(x-2) = 3x - 2(x-1) + \dfrac{16}{3}$

$3\left(\dfrac{1}{3}\right)(x-2) = 3(3x) - 3(2)(x-1) + 3\left(\dfrac{16}{3}\right)$

$x - 2 = 9x - 6x + 6 + 16$

$x - 2 = 3x + 22$

$-2x = 24$

$x = -12$

37. $\dfrac{1}{2}x + \dfrac{1}{9}x = \dfrac{3x-2}{12}$

$36\left(\dfrac{1}{2}x\right) + 36\left(\dfrac{1}{9}x\right) = 36\left(\dfrac{3x-2}{12}\right)$

$18x + 4x = 9x - 6$

$22x = 9x - 6$

$13x = -6$

$x = -\dfrac{6}{13}$

39. $0.8(x-3) = -5(2.1x + 0.4)$

$0.8x - 2.4 = -10.5x - 2$

$10(0.8x) - 10(2.4) = 10(-10.5x) - 10(2)$

$8x - 24 = -105x - 20$

$113x = 4$

$x = \dfrac{4}{113}$

41. $\dfrac{1}{5}x + \dfrac{2}{3} + \dfrac{1}{15} = \dfrac{4}{3}x - \dfrac{7}{15}$

$15\left(\dfrac{1}{5}x\right) + 15\left(\dfrac{2}{3}\right) + 15\left(\dfrac{1}{15}\right) = 15\left(\dfrac{4}{3}x\right) - 15\left(\dfrac{7}{15}\right)$

$3x + 10 + 1 = 20x - 7$

$-17x = -18$

$x = \dfrac{18}{17}$

43. $7x + 2[(x+2)+3] + 2(x+2) = 2[2(x+2)+2x] + 8$

$7x + 2(x+5) + 2(x+2) = 2(2x+4+2x) + 8$

$7x + 2x + 10 + 2x + 4 = 8x + 8 + 8$

$11x + 14 = 8x + 16$

$3x = 2$

$x = \dfrac{2}{3}$

Cumulative Review Problems

45. $\dfrac{3}{7} + 1\dfrac{5}{10}$

$= \dfrac{3}{7} + \dfrac{15}{10}$

$= \dfrac{30}{70} + \dfrac{105}{70}$

$= \dfrac{135}{70} = 1\dfrac{65}{70} = 1\dfrac{13}{14}$

47. $\left(2\dfrac{1}{5}\right)\left(6\dfrac{1}{8}\right) = \left(\dfrac{11}{5}\right)\left(\dfrac{49}{8}\right)$

$= \dfrac{539}{40}$

$= 13\dfrac{19}{40}$

Exercises 2.5

1. $d = rt$

$\dfrac{d}{t} = \dfrac{rt}{t}$

$\dfrac{d}{t} = r$

$d = 450$ mi, $t = 9h$

$r = \dfrac{450}{9} = 50$

rate $= 50$ mi / h

3. (a) $A = \dfrac{1}{2}bh$

$\dfrac{2A}{h} = \dfrac{2\left(\dfrac{1}{2}bh\right)}{h} = b$

$A = 60\ m^2,\ h = 12\ m$

$b = \dfrac{2(60)}{12} = 10$

base $= 10\ m$

3. (b) $A = \dfrac{1}{2}bh$

$$\dfrac{2A}{b} = \dfrac{2\left(\dfrac{1}{2}bh\right)}{b}$$

$$\dfrac{2A}{b} = h$$

$A = 88\ m^2, b = 11\ m$

$$h = \dfrac{2(88)}{11} = 16$$

height $= 16\ m$

5. (a) $3x - 5y = 15$

$$3x + (-3x) - 5y = -3x + 15$$

$$-5y = -3x + 15$$

$$\dfrac{-5y}{-5} = \dfrac{-3x + 15}{-5}$$

$$y = \dfrac{-3x + 15}{-5} \text{ or } y = \dfrac{3}{5}x - 3$$

(b) $x = -5$

$$y = \dfrac{3}{5}(-5) - 3 = -6$$

7. $A = \dfrac{1}{2}bh$

$$\dfrac{2A}{b} = \dfrac{2\left(\dfrac{1}{2}bh\right)}{b}$$

$$\dfrac{2A}{b} = h$$

9. $I = Ptr$

$$\dfrac{I}{Pt} = \dfrac{Ptr}{Pt}$$

$$\dfrac{I}{Pt} = r$$

11. $y = mx + b$

$$y + (-mx) = mx + (-mx) + b$$

$$y - mx = b$$

13. $A = P(1 + rt)$

$$A = P + Ptr$$

$$A + (-P) = P + (-P) + Ptr$$

$$A - P = Ptr$$

$$\dfrac{A - P}{Pt} = \dfrac{Ptr}{Pt}$$

$$\dfrac{A - P}{Pt} = r$$

15. $5x + 9y = -18$

$$5x + (-5x) + 9y = -5x - 18$$

$$9y = -5x - 18$$

$$\dfrac{9y}{9} = \dfrac{-5x - 18}{9}$$

$$y = -\dfrac{5}{9}x - 2$$

17. $y = \dfrac{6}{7}x - 12$

$$7y = 7\left(\dfrac{6}{7}x\right) - 7(12)$$

$$7y = 6x - 84$$

$$7y + 84 = 6x - 84 + 84$$

$$7y + 84 = 6x$$

$$\dfrac{7y + 84}{6} = \dfrac{6x}{6}$$

$$\dfrac{7}{6}y + 14 = x$$

19. $ax + by = c$

$$ax + by + (-by) = -by + c$$

$$ax = -by + c$$

$$\dfrac{ax}{a} = \dfrac{-by + c}{a}$$

$$x = \dfrac{-by + c}{a} \text{ or } x = -\dfrac{b}{a}y + \dfrac{c}{a}$$

21. $s = 4\pi r^2$

$$\dfrac{s}{4\pi} = \dfrac{4\pi r^2}{4\pi}$$

$$\dfrac{s}{4\pi} = r^2$$

23. $E = mc^2$

$$\frac{E}{c^2} = \frac{mc^2}{c^2}$$

$$\frac{E}{c^2} = m$$

25. $S = \frac{1}{2}gt^2$

$$2S = 2\left(\frac{1}{2}gt^2\right)$$

$$2S = gt^2$$

$$\frac{2S}{g} = \frac{gt^2}{g}$$

$$\frac{2S}{g} = t^2$$

27. $E = IR$

$$\frac{E}{R} = \frac{IR}{R}$$

$$\frac{E}{R} = I$$

29. $V = LWH$

$$\frac{V}{LW} = \frac{LWH}{LW}$$

$$\frac{V}{LW} = H$$

31. $V = \frac{1}{3}Bh$

$$3V = 3\left(\frac{1}{3}Bh\right)$$

$$3V = Bh$$

$$\frac{3V}{h} = \frac{Bh}{h}$$

$$\frac{3V}{h} = B$$

33. $P = 2L + 2W$

$$P + (-2W) = 2L + 2W + (-2W)$$

$$P - 2W = 2L$$

$$\frac{P - 2W}{2} = \frac{2L}{2}$$

$$\frac{P - 2W}{2} = L \text{ or } L = \frac{P}{2} - W$$

35. $c^2 = a^2 + b^2$

$$c^2 + (-a^2) = a^2 + (-a^2) + b^2$$

$$c^2 - a^2 = b^2$$

37. $C = \frac{5}{9}(F - 32)$

$$\frac{9}{5}C = \frac{9}{5}\left(\frac{5}{9}\right)(F - 32)$$

$$\frac{9}{5}C = F - 32$$

$$\frac{9}{5}C + 32 = F - 32 + 32$$

$$\frac{9}{5}C + 32 = F$$

39. $d = 16t^2 + vt$

$$d + (-16t^2) = 16t^2 + (-16t^2) + vt$$

$$d - 16t^2 = vt$$

$$\frac{d - 16t^2}{t} = \frac{vt}{t}$$

$$\frac{d - 16t^2}{t} = v \text{ or } v = \frac{d}{t} - 16t$$

To Think About

41. $I = \text{Pr}t$

If $t \Rightarrow 2t$, $I \Rightarrow \text{Pr}(2t) = 2I$

I doubles if t doubles.

43. $A = \pi r^2$

If $r \Rightarrow 2r$, $A \Rightarrow \pi(2r)^2 = 4\pi r^2$

A is quadrupled if r doubles.

Cumulative Review Problems

45. 12% of $260 = 0.12(260) = 31.2$

47. $\frac{12}{30} = 0.40 = 40\%$

Putting Your Skills to Work

1. $d = rt$, $r = 1100\frac{\text{ft}}{s}$, $t = 1.4s$

$$d = (1100)(1.4) = 1540 \text{ ft}$$

3. You: $d = rt = 1100(3.0) = 3300$ ft

   Friend: $d = rt = 1100(3.5) = 3850$ ft

   She is farther from the storm

   Distance apart $= 3300 + 3850 = 7150$ ft

5. Because lighting travels about 1 trillion times faster than sound, we do not significantly affect our estimate of distance by ignoring the time for the light to reach our eyes.

Exercises 2.6

1. $5 > -6$ is equivalent to $-6 < 5$

3. $9 > -3$

5. $-4 < -2$

7. $\dfrac{2}{3} < \dfrac{3}{4}$

9. $-1.2 < 2.1$

11. $-\dfrac{13}{3} < -4$

13. $-6.9 > -7.2$

15. $\dfrac{123}{4986} \approx 0.024669$

    $0.0247 > \dfrac{123}{4986}$

17. $x \geq -6$

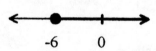

19. $x < 3$

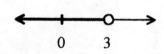

21. $x > \dfrac{3}{4}$

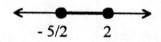

23. $x \leq -3.6$

25. $x > 35$

27. $x \geq -1$

29. $x < -20$

31. $x > 6.5$

33. $c < 56$

35. $h \geq 37$

37. $x \leq 2, \quad x > -3, \quad x < \dfrac{5}{2}, \quad x \geq -\dfrac{5}{2}$

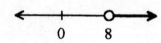

Cumulative Review Problems

39. 16% of $38 = 0.16(38) = 6.08$

41. $\dfrac{16}{800} = 0.02 = 2\%$

Exercises 2.7

1. $x - 5 > 3$

   $x > 8$

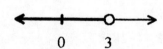

3. $4x < 12$

   $x < 3$

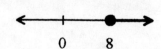

5. $\dfrac{1}{2}x \geq 4$

   $x \geq 8$

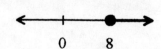

7. $2x - 3 < 4$

$2x < 7$

$x < \dfrac{7}{2}$

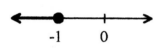

      0   7/2

9. $6 - 5x \geq 3x + 14$

$-8x \geq 8$

$x \leq -1$

      -1   0

11. $\dfrac{5x}{6} - 5 > \dfrac{x}{6} - 9$

$5x - 30 > x - 54$

$4x > -24$

$x > -6$

     -6   0

13. $3(x + 2) \leq 2x + 4$

$3x + 6 \leq 2x + 4$

$x \leq -2$

     -2   0

15. $\qquad 5 > 3$

$5 + (-2) > 3 + (-2)$

$\qquad 3 > 1$

Adding any number to both sides of an inequality does not reverse the direction.

17. $2x - 5 < 5x - 11$

$-3x < -6$

$x > 2$

19. $6x - 2 \geq 4x + 6$

$2x \geq 8$

$x \geq 4$

21. $0.3(x - 1) < 0.1x - 0.5$

$0.3x - 0.3 < 0.1x - 0.5$

$3x - 3 < x - 5$

$2x < -2$

$x < -1$

23. $5 - 2(3 - x) \leq 2(2x + 5) + 1$

$5 - 6 + 2x \leq 4x + 10 + 1$

$-1 + 2x \leq 4x + 11$

$-2x \leq 12$

$x \geq -6$

25. $2x - \left(\dfrac{x}{2} - 6\right) \geq 1 + \dfrac{2x}{3}$

$2x - \dfrac{x}{2} + 6 \geq 1 + \dfrac{2x}{3}$

$12x - 3x + 36 \geq 6 + 4x$

$9x + 36 \geq 6 + 4x$

$5x \geq -30$

$x \geq -6$

27. $3(0.3 + 0.1x) + 0.1 < 0.5(x + 2)$

$0.9 + 0.3x + 0.1 < 0.5x + 1$

$9 + 3x + 1 < 5x + 10$

$3x + 10 < 5x + 10$

$-2x < 0$

$x > 0$

29. $\dfrac{1}{6} - \dfrac{1}{2}(3x + 2) < \dfrac{1}{3}\left(x - \dfrac{1}{2}\right)$

$1 - 3(3x + 2) < 2\left(x - \dfrac{1}{2}\right)$

$1 - 9x - 6 < 2x - 1$

$-9x - 5 < 2x - 1$

$-11x < 4$

$x > -\dfrac{4}{11}$

31. $1.96x - 2.58 < 9.36x + 8.21$

$-7.40x < 10.79$

$x > -1.46$

33. $\dfrac{75 + 83 + 86 + x}{4} \geq 80$

$$\dfrac{244 + x}{4} \geq 80$$

$$244 + x \geq 320$$

$$x \geq 76$$

Must get a 76 or higher.

35. $P = 12.5n - 300,000$

a. $0 = 12.5n - 300,000$

$n = 24,000$

Manufacture 24,000 chips.

b. $12.5n - 300,000 \geq 650,000$

$12.5n \geq 950,000$

$n \geq 76,000$

Manufacture at least 76,000 chips.

$12.5n - 300,000 \geq 470,000$

$12.5n \geq 770,000$

$n \geq 61,600$

Manufacture at least 61,600 chips.

Putting Your Skills to Work

1. Air Fare $\geq 490 + 310$

≥ 800

The least is $800

3. Air Fare $\geq 500 + 310$

≥ 810

The least is $810

Air Fare $\leq 730 + 570$

≤ 1300

The most is $1300

Chapter 2 - Review Problems

1. $\qquad 5x + 20 = 3x$

$5x + (-3x) + 20 = 3x + (-3x)$

$2x + 20 = 0$

$2x + 20 + (-20) = +(-20)$

$2x = -20$

$\dfrac{2x}{2} = \dfrac{-20}{2}$

$x = -10$

3. $\qquad 7(x - 4) = x + 2$

$7x - 28 = x + 2$

$7x + (-x) - 28 = x + (-x) + 2$

$6x - 28 = 2$

$6x - 28 + 28 = 2 + 28$

$6x = 30$

$\dfrac{6x}{6} = \dfrac{30}{6}$

$x = 5$

5. $4x - 3(x + 2) = 4$

$4x - 3x - 6 = 4$

$x - 6 = 4$

$x - 6 + 6 = 4 + 6$

$x = 10$

7. $x - (0.5x + 2.6) = 17.6$

$x - 0.5x - 2.6 = 17.6$

$0.5x - 2.6 = 17.6$

$10(0.5x) - 10(2.6) = 10(17.6)$

$5x - 26 = 176$

$5x - 26 + 26 = 176 + 26$

$5x = 202$

$\dfrac{5x}{5} = \dfrac{202}{5}$

$x = 40.4$

9.
$$3(x-2)=-4(5+x)$$
$$3x-6=-20-4x$$
$$3x+4x-6=-20-4x+4x$$
$$7x-6=-20$$
$$7x-6+6=-20+6$$
$$7x=-14$$
$$\frac{7x}{7}=\frac{-14}{7}$$
$$x=-2$$

11.
$$x+37=26$$
$$x+37+(-37)=26+(-37)$$
$$x=-11$$

13.
$$3(x-3)=13x+21$$
$$3x-9=13x+21$$
$$3x+(-3x)-9=13x+(-3x)+21$$
$$-9=10x+21$$
$$-21-9=10x+21+(-21)$$
$$-30=10x$$
$$\frac{-30}{10}=\frac{10x}{10}$$
$$-3=x$$

15.
$$24-3x=4(x-1)$$
$$24-3x=4x-4$$
$$24-3x+3x=4x+3x-4$$
$$24=7x-4$$
$$24+4=7x-4+4$$
$$28=7x$$
$$\frac{28}{7}=\frac{7x}{7}$$
$$4=x$$

17.
$$36=9x-(3x-18)$$
$$36=9x-3x+18$$
$$36=6x+18$$
$$36+(-18)=6x+18+(-18)$$
$$18=6x$$
$$\frac{18}{6}=\frac{6x}{6}$$
$$3=x$$

19.
$$2(3-x)=1-(x-2)$$
$$6-2x=1-x+2$$
$$6-2x+x=3-x+x$$
$$6-x=3$$
$$6+(-6)-x=3+(-6)$$
$$-x=-3$$
$$\frac{-x}{-1}=\frac{-3}{-1}$$
$$x=3$$

21.
$$0.9x+3=0.4x+1.5$$
$$10(0.9x)+10(3)=10(0.4x)+10(1.5)$$
$$9x+30=4x+15$$
$$9x+(-4x)+30=4x+(-4x)+15$$
$$5x+30=15$$
$$54x+30+(-30)=15+(-30)$$
$$5x=-15$$
$$\frac{5x}{5}=\frac{-15}{5}$$
$$x=-3$$

23.
$$8(3x+5)-10=9(x-2)+13$$
$$24x+40-10=9x-18+13$$
$$24x+30=9x-5$$
$$24x+(-9x)+30=9x+(-9x)-5$$
$$15x+30=-5$$
$$15x+30+(-30)=-5+(-30)$$
$$15x=-35$$
$$\frac{15x}{15}=\frac{-35}{15}$$
$$x=\frac{-7}{3}$$

25.
$$3=2x+5-3(x-1)$$
$$3=2x+5-3x+3$$
$$3=-x+8$$
$$3+(-8)=-x+8+(-8)$$
$$-5=-x$$
$$\frac{-5}{-1}=\frac{-x}{-1}$$
$$5=x$$

27. $2(5x-1)-7 = 3(x-1)+5-4x$

$\quad 10x-2-7 = 3x-3+5-4x$

$\quad 10x-9 = -x+2$

$\quad 10x+x-9 = -x+x+2$

$\quad 11x-9 = 2$

$\quad 11x-9+9 = 2+9$

$\quad 11x = 11$

$\quad \dfrac{11x}{11} = \dfrac{11}{11}$

$\quad x = 1$

29. $\quad 1 = \dfrac{5x}{6} - \dfrac{2x}{3}$

$\quad 6(1) = 6\left(\dfrac{5x}{6}\right) - 6\left(\dfrac{2x}{3}\right)$

$\quad 6 = 5x - 4x$

$\quad 6 = x$

31. $\quad \dfrac{7x-3}{2} - 4 = \dfrac{5x+1}{3}$

$\quad 6\left(\dfrac{7x-3}{2}\right) - 6(4) = 6\left(\dfrac{5x+1}{3}\right)$

$\quad 3(7x-3) - 24 = 2(5x+1)$

$\quad 21x-9-24 = 10x+2$

$\quad 21x-33 = 10x+2$

$\quad 21x+(-10x)-33 = 10x+(-10x)+2$

$\quad 11x-33 = 2$

$\quad 11x-33+33 = 2+33$

$\quad 11x = 35$

$\quad \dfrac{11x}{11} = \dfrac{35}{11}$

$\quad x = \dfrac{35}{11}$

33. $\quad \dfrac{-3}{2}(x+5) = 1-x$

$\quad 2\left(\dfrac{-3}{2}\right)(x+5) = 2(1-x)$

$\quad -3x-15 = 2-2x$

$\quad -3x+3x-15 = 2-2x+3x$

$\quad -15 = 2+x$

$\quad -15+(-2) = 2+(-2)+x$

$\quad -17 = x$

35. $\quad \dfrac{1}{3}(x-2) = \dfrac{x}{4} + 2$

$\quad 12\left(\dfrac{1}{3}\right)(x-2) = 12\left(\dfrac{x}{4}\right) + 12(2)$

$\quad 4(x-2) = 3x+24$

$\quad 4x-8 = 3x+24$

$\quad 4x+(-3x)-8 = 3x+(-3x)+24$

$\quad x-8 = 24$

$\quad x-8+8 = 24+8$

$\quad x = 32$

37. $\quad \dfrac{4}{5} + \dfrac{1}{2}x = \dfrac{1}{5}x + \dfrac{1}{2}$

$\quad 10\left(\dfrac{4}{5}\right) + 10\left(\dfrac{1}{2}x\right) = 10\left(\dfrac{1}{5}x\right) + 10\left(\dfrac{1}{2}\right)$

$\quad 8+5x = 2x+5$

$\quad 5x+(-2x)+8 = 2x+(-2x)+5$

$\quad 3x+8 = 5$

$\quad 3x+8+(-8) = 5+(-8)$

$\quad 3x = -3$

$\quad \dfrac{3x}{3} = \dfrac{-3}{3}$

$\quad x = -1$

39. $\quad \dfrac{10}{3} - \dfrac{5}{3}x + x = \dfrac{2}{9} + \dfrac{1}{9}x$

$\quad 9\left(\dfrac{10}{3}\right) - 9\left(\dfrac{5}{3}x\right) + 9x = 9\left(\dfrac{2}{9}\right) + 9\left(\dfrac{1}{9}x\right)$

$\quad 30-15x+9x = 2+x$

$\quad 30-6x = 2+x$

$\quad 30-6x+6x = 2+x+6x$

$\quad 30 = 2+7x$

$\quad 30+(-2) = 2+(-2)+7x$

$\quad 28 = 7x$

$\quad \dfrac{28}{7} = \dfrac{7x}{7}$

$\quad 4 = x$

41.
$$\frac{1}{2} + \frac{5}{4}x = \frac{2}{5}x - \frac{1}{10} + 4$$
$$20\left(\frac{1}{2}\right) + 20\left(\frac{5}{4}x\right) = 20\left(\frac{2}{5}x\right) - 20\left(\frac{1}{10}\right) + 20(4)$$
$$10 + 25x = 8x - 2 + 80$$
$$10 + 25x = 8x + 78$$
$$10 + 25x + (-8x) = 8x + (-8x) + 78$$
$$10 + 17x = 78$$
$$10 + (-10) + 17x = 78 + (-10)$$
$$17x = 68$$
$$\frac{17x}{17} = \frac{68}{17}$$
$$x = 4$$

43.
$$\frac{1}{6}x - \frac{2}{3} = \frac{1}{3}(x - 4)$$
$$6\left(\frac{1}{6}x\right) - 6\left(\frac{2}{3}\right) = 6\left(\frac{1}{3}\right)(x - 4)$$
$$x - 4 = 2x - 8$$
$$x + (-x) - 4 = 2x + (-x) - 8$$
$$-4 = x - 8$$
$$-4 + 8 = x - 8 + 8$$
$$4 = x$$

45.
$$\frac{7}{12}(x - 3) = -\frac{1}{3}x + 1$$
$$12\left(\frac{7}{12}\right)(x - 3) = -12\left(\frac{1}{3}x\right) + 12(1)$$
$$7x - 21 = -4x + 12$$
$$7x + 4x - 21 = -4x + 4x + 12$$
$$11x - 21 = 12$$
$$11x - 21 + 21 = 12 + 21$$
$$11x = 33$$
$$\frac{11x}{11} = \frac{33}{11}$$
$$x = 3$$

47.
$$\frac{1}{7}(x + 5) - \frac{6}{14} = \frac{1}{2}(x + 3)$$
$$14\left(\frac{1}{7}\right)(x + 5) - 14\left(\frac{6}{14}\right) = 14\left(\frac{1}{2}\right)(x + 3)$$
$$2x + 10 - 6 = 7x + 21$$
$$2x + 4 = 7x + 21$$
$$2x + (-2x) + 4 = 7x + (-2x) + 21$$
$$4 = 5x + 21$$
$$4 + (-21) = 5x + 21 + (-21)$$
$$-17 = 5x$$
$$\frac{-17}{5} = \frac{5x}{5}$$
$$-\frac{17}{5} = x$$

49.
$$-\frac{1}{3}(2x - 6) = \frac{2}{3}(3 + x)$$
$$3\left(-\frac{1}{3}\right)(2x - 6) = 3\left(\frac{2}{3}\right)(3 + x)$$
$$-2x + 6 = 6 + 2x$$
$$-2x + 6 = 6 + 2x + 2x$$
$$6 = 6 + 4x$$
$$6 + (-6) = 6 + (-6) + 4x$$
$$0 = 4x$$
$$\frac{0}{4} = \frac{4x}{4}$$
$$0 = x$$

51.
$$3x - y = 10$$
$$3x + (-3x) - y = -3x + 10$$
$$-y = -3x + 10$$
$$\frac{-y}{-1} = \frac{-3x + 10}{-1}$$
$$y = 3x - 10$$

53.
$$A = P(1 + rt)$$
$$A = P + \mathrm{Pr}\,t$$
$$A + (-P) = P + (-P) + \mathrm{Pr}\,t$$
$$A - P = \mathrm{Pr}\,t$$
$$\frac{A - P}{Pt} = \frac{\mathrm{Pr}\,t}{Pt}$$
$$\frac{A - P}{Pt} = r$$

55.
$$H = \frac{1}{3}(a + 2p + 3)$$
$$3H = 3\left(\frac{1}{3}\right)(a + 2p + 3)$$
$$3H = a + 2p + 3$$
$$3H + (-a) + (-3) = a + (-a) + 2p + 3 + (-3)$$
$$3H - a - 3 = 2p$$
$$\frac{3H - a - 3}{2} = \frac{2p}{2}$$
$$\frac{3H - a - 3}{2} = p$$

57. (a) $5x - 3y = 12$
$$5x + (-5x) - 3y = -5x + 12$$
$$-3y = -5x + 12$$
$$\frac{-3y}{-3} = \frac{-5x + 12}{-3}$$
$$y = \frac{5}{3}x - 4$$

(b) $x = 9$
$$y = \frac{5}{3}(9) - 4 = 15 - 4 = 11$$

59. $7 - 2x \geq 4x - 5$
$$-6x \geq -12$$
$$x \leq 2$$

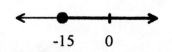

61. $2x - 3 + x > 5(x + 1)$
$$3x - 3 > 5x + 5$$
$$-2x > 8$$
$$x < -4$$

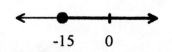

63. $4x \geq 2(12 - 2x)$
$$4x \geq 24 - 4x$$
$$8x \geq 24$$
$$x \geq 3$$

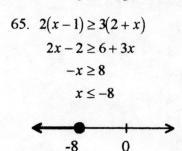

65. $2(x - 1) \geq 3(2 + x)$
$$2x - 2 \geq 6 + 3x$$
$$-x \geq 8$$
$$x \leq -8$$

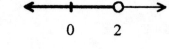

67. $4x - 14 < 4 - 2(3x - 1)$
$$4x - 14 < 4 - 6x + 2$$
$$4x - 14 < 6 - 6x$$
$$10x < 20$$
$$x < 2$$

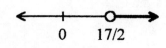

69. $\frac{1}{2}(2x + 3) > 10$
$$2x + 3 > 20$$
$$2x > 17$$
$$x > \frac{17}{2}$$

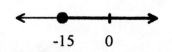

71. $4(2 - x) - (-5x + 1) \geq -8$
$$8 - 4x + 5x - 1 \geq -8$$
$$x + 7 \geq -8$$
$$x \geq -15$$

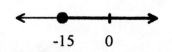

73. $70n \le 3220$

$$n \le \frac{3220}{70}$$

$$n \le 46$$

Test - Chapter 2

1. $2x + 0.8 = 5.0$

$2x = 4.2$

$x = 2.1$

3. $2(2y - 3) = 4(2y + 2)$

$4y - 6 = 8y + 8$

$-4y = 14$

$$y = -\frac{7}{2}$$

5. $2(6 - x) = 3x + 2$

$12 - 2x = 3x + 2$

$10 = 5x$

$2 = x$

7. $\dfrac{2y}{3} + \dfrac{1}{5} - \dfrac{3y}{5} + \dfrac{1}{3} = 1$

$10y + 3 - 9y + 5 = 15$

$y + 8 = 15$

$y = 7$

9. $5(20 - x) + 10x = 165$

$100 - 5x + 10x = 165$

$100 + 5x = 165$

$5x = 65$

$x = 13$

11. $-2(2 - 3x) = 76 - 2x$

$-4 + 6x = 76 - 2x$

$8x = 80$

$x = 10$

13. $2x - 3 = 12 - 6x + 3(2x + 3)$

$2x - 3 = 12 - 6x + 6x + 9$

$2x - 3 = 21$

$2x = 24$

$x = 12$

15. $\dfrac{3}{5}x + \dfrac{7}{10} = \dfrac{1}{3}x + \dfrac{3}{2}$

$18x + 21 = 10x + 45$

$8x = 24$

$x = 3$

17. $\dfrac{1}{3}(7x - 1) + \dfrac{1}{4}(2 - 5x) = \dfrac{1}{3}(5 - 3x)$

$4(7x - 1) + 3(2 - 5x) = 4(5 - 3x)$

$28x - 4 + 6 - 15x = 20 - 12x$

$13x + 2 = 20 - 12x$

$25x = 18$

$$x = \frac{18}{25}$$

19. $\dfrac{2w}{3} = 4 - \dfrac{1}{2}(x + 6)$

$4w = 24 - 3(x + 6)$

$4w = 24 - 3x - 18$

$4w = 6 - 3x$

$$w = \frac{6 - 3x}{4}$$

21. $5ax(2 - y) = 3axy + 5$

$10ax - 5axy = 3axy + 5$

$10ax = 8axy + 5$

$10ax - 5 = 8axy$

$$\frac{10ax - 5}{8ax} = y$$

23. $2 - 7(x + 1) - 5(x + 2) < 0$

$2 - 7x - 7 - 5x - 10 < 0$

$-12x - 15 < 0$

$-12x < 15$

$$x > -\frac{5}{4}$$

$-5/4 \qquad 0$

25. $\dfrac{1}{4}x + \dfrac{1}{16} \le \dfrac{1}{8}(7x - 2)$

$4x + 1 \le 14x - 4$

$-10x \le -5$

$$x \ge \frac{1}{2}$$

Cumulative Test - Chapters 0 - 2

1. $\dfrac{6}{7} - \dfrac{2}{3} = \dfrac{18}{21} - \dfrac{14}{21} = \dfrac{4}{21}$

3. $3\dfrac{1}{5} \div 1\dfrac{1}{2} = \dfrac{16}{5} \cdot \dfrac{2}{3} = \dfrac{32}{15} = 2\dfrac{2}{15}$

5. $\begin{array}{r} 0.12 \\ 1.2\overline{)0.144} \\ \underline{12} \\ 24 \\ \underline{24} \end{array}$

7. $(-3)(-5)(-1)(2)(-1) = (3)(5)(1)(2)(1) = 30$

9. $\left(5x^2y^3\right)^2 = 5^2\left(x^2\right)\left(y^3\right)^2 = 25x^4y^6$

11. $3(5 - x) = 2x - 10$
    $15 - 3x = 2x - 10$
    $25 = 5x$
    $5 = x$

13. $\dfrac{2y}{3} - \dfrac{1}{4} = \dfrac{1}{6} + \dfrac{y}{4}$
    $8y - 3 = 2 + 3y$
    $5y = 5$
    $y = 1$

15. $H = \dfrac{2}{3}(b + 4a)$
    $\dfrac{3H}{2} = b + 4a$
    $\dfrac{3H}{2} - 4a = b$ or $b = \dfrac{3H - 8a}{2}$

17. $A = \dfrac{ha}{2} + \dfrac{hb}{2}$
    $2A = ha + hb$
    $2A - hb = ha$
    $\dfrac{2A - hb}{h} = a$ or $a = \dfrac{2A}{h} - b$

19. $\dfrac{1}{2}(x - 5) \geq x - 4$
    $x - 5 \geq 2x - 8$
    $-x \geq -3$
    $x \leq 3$

21. $x + \dfrac{5}{9} \leq \dfrac{1}{3} + \dfrac{7}{9}x$
    $9x + 5 \leq 3 + 7x$
    $2x \leq -2$
    $x \leq -1$

Practice Quiz: Sections 2.1 - 2.3

Solve for x.

1. $x - 7 = 21$

2. $3x = 8$

3. $-\dfrac{1}{5}x = 20$

4. $3x - 6 = 12$

5. $6 - 4x = 5x + 2$

Practice Quiz: Sections 2.4 - 2.7

In Problems 1 - 3 solve for x.

1. $\dfrac{1}{4}x + \dfrac{2}{3} = \dfrac{1}{6}x - 2$

2. $3(5x - 3) = \dfrac{1}{2}(x - 2) + 4$

3. $3x - 4y = 8$

In Problems 4-5 solve and graph the inequalities.

4. $x + 8 \leq 5$

5. $\dfrac{1}{2}(x - 3) > 4 - \dfrac{1}{3}x$

<u>Answers to Practice Quiz</u>

Sections 2.1 - 2.3

1. $x = 28$

2. $x = \dfrac{8}{3}$

3. $x = -100$

4. $x = 6$

5. $x = \dfrac{4}{9}$

Sections 2.4 - 2.7

1. $x = -32$

2. $x = \dfrac{24}{29}$

3. $x = \dfrac{4y + 8}{3}$

4. $x \leq -3$

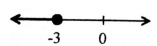

5. $x > \dfrac{33}{5}$

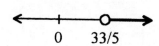

Pretest Chapter 3

1. $2x - 30$

3. $x =$ the number
$$7x + 9 = 44$$
$$7x = 35$$
$$x = 5$$
The number is 5

5. $x =$ length of second side,
$2x - 3 =$ length of first side,
$2x - 3 + 4 =$ length of third side
$$x + 2x - 3 + 2x - 3 + 4 = 41$$
$$5x = 43$$
$$x = 8.6, \text{ also}$$
$$2x - 3 = 14.2, \text{ and}$$
$$2x - 3 + 4 = 18.2$$
The lengths are 14.2 cm, 8.6 cm, 18.2 cm.

7. $w =$ width,
$2w + 13 =$ length
$$P = 2L + 2W$$
$$74 = 2(2w + 13) + 2w$$
$$74 = 4w + 26 + 2w$$
$$48 = 6w$$
$$8 = w, \text{ and}$$
$$2w + 13 = 29$$
Width $= 8$ m and Length $= 29$ m

9. $x =$ original population,
$0.11x =$ population growth
$$x + 0.11x = 24{,}420$$
$$1.11x = 24{,}420$$
$$x = 22{,}000$$
The population was 22,000

11. $b = 16, h = 9$
$$A = \frac{1}{2}bh = \frac{1}{2}(16)(9) = 72 \text{ in}^2$$

13. $r = 4$
$$C = 2\pi r = 2(3.14)(4) = 25.12 \text{ cm}$$

15. $h \leq 72$

17. $n < 15$

19. $x =$ amount of sales
$$0.03x > 900$$
$$x > 30{,}000$$
He must sell more than \$30,000

Exercises 3.1

1. $x + 7$

3. $x - 5$

5. $\dfrac{x}{3}$

7. $3x$

9. $\dfrac{x}{8}$

11. $2x + 5$

13. $\dfrac{2}{7}x - 3$

15. $5(x + 12)$

17. $\dfrac{2}{3}(x - 6)$

19. $\dfrac{x}{2} + 3x$

21. $4x - \dfrac{x}{2}$

23. $b =$ no. of hours Barbara works
$b + 15 =$ no. of hours Alicia works

25. $c =$ cost of computer
$c - 390 =$ cost of printer

27. $w =$ width of rectangle
$3w + 3 =$ length of rectangle

29. $x =$ attendance Saturday
$x + 1600 =$ attendance Friday
$x - 783 =$ attendance Thursday

31. t = measure of 3rd angle

$t + 19$ = measure of 1st angle

$3t$ = measure of 2nd angle

33. x = attendance Thursday

$3x - 200$ = attendance Friday

35. x = measure of 1st angle

$x - 6$ = measure of 2nd angle

$2x$ = measure of 3rd angle

37. x = sales in March

$\dfrac{3}{4}x$ = sales in April

$\dfrac{5}{6}x$ = sales in February

39. x = last year's profit

$3x - 400 - 0.5x$ = this year's profit

Cumulative Review Problems

41. $2x + 2(3x - 4) = 72$

$2x + 6x - 8 = 72$

$8x = 80$

$x = 10$

43. $6(w - 1) - 3(2 + w) = 9$

$6w - 6 - 6 - 3w = 9$

$3w = 21$

$w = 7$

Exercises 3.2

1. x = the number

$125 + x = 170$

$x = 45$

Number is 45

3. x = the number

$\dfrac{2}{5}x = 288$

$x = 720$

Number is 720

5. x = the number

$x + 15 = 305$

$x = 290$

Number is 290

7. x = the number

$2x + 3 = 81$

$2x = 78$

$x = 39$

Number is 39

9. x = original number

$\dfrac{1}{3}x - 6 = 11$

$\dfrac{1}{3}x = 17$

$x = 51$

Original number is 51

11. x = original number

$4x - 30 = 7x$

$-30 = 3x$

$-10 = x$

Original number is -10

13. x = one number, $x + 12$ = the other

$x + (x + 12) = 156$

$2x + 12 = 156$

$2x = 144$

$x = 72$, and

$x + 12 = 84$

Numbers are 72, 84

15. x = one number,

$2x - 3$ = the other

$x + (2x - 3) = 48$

$3x = 51$

$x = 17$, and

$2x - 3 = 31$

Numbers are 17, 31

17. x = original number

$$\frac{x}{4} + \frac{x}{8} + \frac{x}{5} = 46$$
$$10x + 5x + 8x = 1840$$
$$23x = 1840$$
$$x = 80$$

Original number is 80

19. x = one number, $x + 41.8632$ = another.

$$x + (x + 41.8632) = 196.0578$$
$$2x = 154.1946$$
$$x = 77.0973, \text{ and}$$
$$x + 41.8632 = 118.9605$$

Numbers are 77.0973, 118.9605

21. x = no. of tables,

     $4x$ = no. of chairs

$$4x = 60$$
$$x = 15$$

Tables in stock: 15

23. n = no. of months

$$320n = 8000$$
$$n = 25$$

Number of months will be 25

25. x = hours of labor

$$45x + 545 = 950$$
$$45x = 405$$
$$x = 9$$

Hours of labor is 9

27. x = number of french fries

$$10x + 315 + 145 = 500$$
$$10x = 40$$
$$x = 4$$

Number of french fries is 4

29. x = number of hours

$$45x = 270$$
$$x = 6$$

Trip took 6 hours

31. x = rate of speed

$$3 = \frac{1}{2}x$$
$$x = 6$$

Rate of speed is 6 mph

33. Tom: $r = 8, t = 2\frac{1}{2}$

$$d = rt = 8\left(2\frac{1}{2}\right) = 20$$

Dave: $r = 10, t = 2\frac{1}{2}$

$$d = rt = 10\left(2\frac{1}{2}\right) = 25$$

Distance apart = $25 - 20 = 5$ miles

35. r = rate

Valley route: $5.5r = 319$
$$r = 58$$

Valley route at 58 mph.

Mountain route: $6r = 312$
$$r = 52$$

Mountain route at 52 mph.

Difference: $58 - 52 = 6$ mph.

37. x = score on last exam

$$\frac{75 + 90 + 85 + 88 + 92 + 73 + 81 + x}{8} = 85$$
$$\frac{584 + x}{8} = 85$$
$$584 + x = 680$$
$$x = 96$$

Score on last exam must be 96

39. x = old station wagon

$$\frac{38 + 21 + x}{3} = 22\frac{2}{3}$$
$$\frac{59 + x}{3} = \frac{68}{3}$$
$$59 + x = 68$$
$$x = 9 \text{ mph}$$

To Think About

41. x = number of cricket chirps

F = Fahrenheit temperature

(a) $F - 40 = \dfrac{x}{4}$

(b) $90 - 40 = \dfrac{x}{4}$

$\qquad 50 = \dfrac{x}{4}$

$\qquad x = 200$ chirps

(c) $F - 40 = \dfrac{148}{4}$

$\qquad F - 40 = 37$

$\qquad F = 77^{0}\,F.$

Cumulative Review Problems

43. $-2a(ab - 3b + 5a) = -2a^2b + 6ab - 10a^2$

45. $5x^2y - 7xy^2 - 8xy - 9x^2y$

$\quad = -4x^2y - 7xy^2 - 8xy$

Exercises 3.3

1. M = Melissa's salary,

$M - 3600$ = Heather's salary.

$M + M - 3600 = 29{,}100$

$\qquad 2M = 32{,}700$

$\qquad M = 16{,}350$, and

$\qquad M - 3600 = 12{,}750$

Melissa earned \$16,350

Heather earned \$12,750

3. x = short length,

$x + 10$ = longer length.

$x + x + 10 = 93$

$\qquad 2x = 83$

$\qquad x = 41.5$, and

$\qquad x + 10 = 51.5$

Short piece = 41.5 meters

Long piece = 51.5 meters

5. x = attendance Tuesday,

$x + 350$ = attendance Monday,

$x + 3450$ = attendance Wednesday.

$x + x + 350 + x + 3450 = 64{,}550$

$\qquad 3x = 60{,}750$

$\qquad x = 20{,}250$, also

$\qquad x + 350 = 20{,}600$, and

$\qquad x + 3450 = 23{,}700$

Monday attendance was 20,600

Tuesday attendance was 20,250

Wednesday attendance was 23,700

7. M = Margaret's loss, $\dfrac{2}{3}M$ = Rita's loss,

$M - 20$ = Tony's loss.

$M + \dfrac{2}{3}M + M - 20 = 156$

$3M + 2M + 3M - 60 = 468$

$\qquad 8M = 528$

$\qquad M = 66$, also

$\qquad \dfrac{2}{3}M = 44$, and

$\qquad M - 20 = 46$

Margaret lost 66 lbs

Rita lost 44 lbs

Tony lost 46 lbs

9. I = current in second transistor,

$\dfrac{1}{2}I$ = current in first, $1.5I$ = current in third.

$I + \dfrac{I}{2} + 1.5I = 0.027$

$\qquad 2I + I + 3I = 0.054$

$\qquad 6I = 0.054$

$\qquad I = 0.009$, also

$\qquad \dfrac{1}{2}I = 0.0045$, and

$\qquad 1.5I = 0.0135$

Current in first = 0.0045 amp

Current in second = 0.009 amp

Current in third = 0.0135 amp

11. w = width, $2w + 3$ = length.

$$P = 2L + 2W$$
$$42 = 2(2w + 3) + 2w$$
$$42 = 4w + 6 + 2w$$
$$36 = 6w$$
$$6 = w, \text{ and}$$
$$2w + 3 = 15$$

Width $= 6$ cm

Length $= 15$ cm

13. w = width, $3w - 20$ = length.

$$P = 2L + 2W$$
$$96 = 2(3w - 20) + 2w$$
$$96 = 6w - 40 + 2w$$
$$136 = 8w$$
$$17 = w, \text{ and}$$
$$3w - 20 = 31$$

Width $= 17$ cm

Length $= 31$ cm

15. x = length of first side, $\frac{1}{2}x$ = second side,

$x - 3$ = third side.

$$x + \frac{1}{2}x + x - 3 = 37$$
$$2x + x + 2x - 6 = 74$$
$$5x = 80$$
$$x = 16, \text{ also}$$
$$\frac{1}{2}x = 8, \text{ and}$$
$$x - 3 = 13$$

First side $= 16$ meters

Second side $= 8$ meters

Third side $= 13$ meters

17. x = length of longest side,

$\frac{1}{2}x + 4$ = shortest, $x - 3$ = second longest.

$$x + x - 3 + \frac{1}{2}x + 4 = 46$$
$$2x + 2x - 6 + x + 8 = 92$$
$$5x = 90$$
$$x = 18, \text{ also}$$
$$x - 3 = 15, \text{ and}$$
$$\frac{1}{2}x + 4 = 13$$

Longest $= 18$ inches

Second longest $= 15$ inches

Shortest $= 13$ inches

19. w = width, $2w + 1.334$ = length.

$$P = 2L + 2W$$
$$49.240 = 2(2w + 1.334) + 2w$$
$$49.240 = 4w + 2.668 + 2w$$
$$46.572 = 6w$$
$$7.762 = w, \text{ and}$$
$$2w + 1.334 = 16.858$$

Width $= 7.762$ cm

Length $= 16.858$ cm

To Think About

21. x = length of side, $x + 2$ = length of new side.

$$4(x + 2) = 5x - 3$$
$$4x + 8 = 5x - 3$$
$$11 = x$$

Dimension of original square is 11 m by 11 m

Cumulative Review Problems

23. $-4x(2x^2 - 3x + 8) = -8x^3 + 12x^2 - 32x$

25. $-7x + 10y - 12x - 8y - 2$
$$= (-7 - 12)x + (10 - 8)y - 2$$
$$= -19x + 2y - 2$$

Exercises 3.4

1. m = miles driven
$$4(18)+.08m = 100.80$$
$$.08m = 28.8$$
$$m = 360$$
She drove 360 miles

3. x = dictionaries sold
$$10(5)+8x = 186$$
$$8x = 136$$
$$x = 17$$
She sold 17 dictionaries

5. x = over time hours
$$6(40)+9x = 303$$
$$9x = 63$$
$$x = 7$$
He needs 7 hours of overtime

7. x = original price
$$0.35x = 84$$
$$x = 240$$
Original price was $240

9. x = crimes last year
$$0.07x = 56$$
$$x = 800$$
Crimes last year were 800

11. x = previous pay, $0.12x$ = pay increase
$$x+0.12x = 16,240$$
$$1.12x = 16,240$$
$$x = 14,500$$
His previous pay was $14,500

13. x = original investment, $0.14x$ = interest earned
$$x+0.14x = 5700$$
$$1.14x = 5700$$
$$x = 5000$$
Original investment was $5000

15. x = amount invested at 12%,
$4000-x$ = amount at 14%.
$$0.12x+0.14(4000-x) = 508$$
$$12x+14(4000-x) = 50,800$$
$$12x+56,000-14x = 50,800$$
$$-2x = -5200$$
$$x = 2600, \text{ and}$$
$$4000-x = 1400$$
Amount at 12% was $2600
Amount at 14% was $1400

17. x = amount invested at 9%,
$6000-x$ = amount at 11%.
$$0.09x+0.11(6000-x) = 624$$
$$9x+11(6000-x) = 62,400$$
$$9x+66,000-11x = 62,400$$
$$-2x = -3600$$
$$x = 1800, \text{ and}$$
$$6000-x = 4200$$
She invested $1800 at 9%
She invested $4200 at 11%

19. x = amount invested at 14%,
$9000-x$ = amount at 9%.
$$0.14x+0.09(9000-x) = 1135$$
$$14x+9(9000-x) = 113,500$$
$$14x+81,000-9x = 113,500$$
$$5x = 32,500$$
$$x = 6500, \text{ and}$$
$$9000-x = 2500$$
They invested $6500 at 14%
and $2500 at 9%

21. x = number of quarters,

    $x - 4$ = number of nickels.

$$0.25x + 0.05(x - 4) = 3.70$$
$$25x + 5(x - 4) = 370$$
$$25x + 5x - 20 = 370$$
$$30x = 390$$
$$x = 13, \text{ and}$$
$$x - 4 = 9$$

She has 13 quarters and 9 nickels

23. x = number of quarters,

    $x - 3$ = number of dimes,

    $2x$ = number of nickels.

$$0.25x + 0.10(x - 3) + 0.05(2x) = 3.75$$
$$25x + 10(x - 3) + 5(2x) = 375$$
$$25x + 10x - 30 + 10x = 375$$
$$45x = 405$$
$$x = 9, \text{ also}$$
$$x - 3 = 6, \text{ and}$$
$$2x = 18$$

He has 9 quarters, 6 dimes, and 18 nickels

25. x = number of nickels,

    $x - 2$ = number of quarters,

    $2x + 1$ = number of half - dollars

$$0.05x + 0.25(x - 2) + 0.5(2x + 1) = 9.10$$
$$5x + 25(x - 2) + 50(2x + 1) = 910$$
$$5x + 25x - 50 + 100x + 50 = 910$$
$$130x = 910$$
$$x = 7, \text{ also}$$
$$x - 2 = 5, \text{ and}$$
$$2x + 1 = 15$$

She has 7 nickels, 5 quarters, and 15 half - dollars

27. x = measure of 1st angle,

    $\frac{1}{2}x$ = measure of 2nd,

    $x + 5$ = measure of 3rd.

$$x + \frac{1}{2}x + x + 5 = 180$$
$$2x + x + 2x + 10 = 360$$
$$5x = 350$$
$$x = 70, \text{ also}$$
$$\frac{1}{2}x = 35, \text{ and}$$
$$x + 5 = 75$$

Angles measure $70^0, 35^0, 75^0$ respectively

29. x = amount invested at 7%,

    $2969 - x$ = amount at 11%.

$$0.07x + 0.11(2969 - x) = 273.47$$
$$7x + 11(2969 - x) = 27{,}347$$
$$7x + 32{,}659 - 11x = 27{,}347$$
$$-4x = -5312$$
$$x = 1328, \text{ and}$$
$$2969 - x = 1641$$

He invested $1328 at 7% and $1641 at 11%

To Think About

31. x = miles driven.

Peerless Agency has the better bargain when the miles driven exceeds the number which makes their total charge equal to that of the Superior Agency.

$$0.24x + 3(35) = 0.16x + 3(41)$$
$$24x + 10{,}500 = 16x + 12{,}300$$
$$8x = 1800$$
$$x = 225$$

They must drive more than 225 miles

Cumulative Review Problems

33. $5(3) + 6 \div (-2) = 15 + (-3) = 12$

35. If $x = -2$ and $y = 3$, then

$$2x^2 + 3xy - 2y^2$$
$$= 2(-2)^2 + 3(-2)(3) - 2(3)^2$$
$$= 8 - 18 - 18 = -28$$

Putting Your Skills to Work

1. $P = \$4000$, $r = 9\%$, $t = 60$ mo.
   Monthly payments $= \$83.04$
   Total Payment $=$ Principal $+$ Interest
   $$60(83.04) = 4000 + I$$
   $$4982.40 - 4000 = I$$
   $$982.40 = I$$
   Total interest paid $= \$982.40$

3. $P = \$2,625$, $r = 10\%$, $t = 72$ mo.
   Monthly payments $= \$48.63$
   Total Payment $=$ Principal $+$ Interest
   $$72(48.63) = 2625 + I$$
   $$3501.36 - 2625 = I$$
   $$876.36 = I$$
   Total interest paid $= \$876.36$

5. $P = \$2,625$, $r = 10\%$, $t = 60$ mo.
   Monthly payments $= \$53.83$
   $$60(53.83) = 2625 + I$$
   $$3229.80 - 2625 = I$$
   $$604.80 = I$$
   Savings $= \$876.36 - 604.80$
   $$= \$271.56$$

Cooperative Group Activity

7. $P = \$5,250$, $r = 8\%$, $t = 48$ mo.
   Monthly payments $= \$128.17$
   Total Payment $= 48(128.17)$
   $$= \$6,152.16$$
   Additional earnings $= 1800(30)$
   $$= \$54,000$$
   Net earnings $= 54,000 - 6,152.16$
   $$= \$47,847.84$$

Exercises 3.5

1. Perimeter is the <u>distance around</u> a plane figure.

3. Area is a measure of the amount of <u>surface</u> in a region.

5. The sum of the interior angles of any triangle is <u>180^0</u>.

7. $h = 14$, $b = 5$
   $$A = bh = (14)(5) = 70 \text{ in}^2$$

9. $A = 80$, $b = 16$
   $$A = \frac{1}{2}bh$$
   $$h = \frac{2A}{b} = \frac{2(80)}{16} = 10 \text{ feet}$$

11. $A = 225$, $w = 15$
    $$A = lw$$
    $$l = \frac{A}{w} = \frac{225}{15} = 15 \text{ in.}$$
    This is a square
    $$P = 4w = 4(15) = 60 \text{ in.}$$

13. $d = 6 \Rightarrow r = \frac{1}{2}d = 3$
    $$A = \pi r^2$$
    $$= (3.14)(3)^2 = 28.26 \ m^2$$

15. $A = 600$, $b_1 = 20$, $b_2 = 30$
    $$A = \frac{1}{2}h(b_1 + b_2)$$
    $$600 = \frac{1}{2}h(20 + 30)$$
    $$1200 = h(50)$$
    $$h = 24 \text{ in}$$

17. Sides $= 12, 27, 31$
    $$P = 12 + 27 + 31$$
    $$= 70 \ m$$

19. $d = 20 \Rightarrow r = \frac{1}{2}d = 10$

$C = 2\pi r$

$= 2(3.14)(10) = 62.8\ m$

21. $P = 27$

$P = 3l$

$27 = 3l$

$l = 9$ in.

23. $l =$ length of equal sides,

$\frac{2}{3}l =$ base, $P = 32$

$P = 2l + \frac{2}{3}l$

$32 = \frac{8}{3}l$

$12 = l$, and

$\frac{2}{3}l = 8$

Equal sides are 12 ft and base $= 8$ ft.

25. Other angle $= 90 - 77$

$= 13^0$

27. $x =$ measure of second angle,

$2x =$ measure of first angle,

$x - 20 =$ measure of third angle

$x + 2x + x - 20 = 180$

$4x = 200$

$x = 50$, also

$2x = 100$, and

$x - 20 = 30$

The angles measure $100^0, 50^0, 30^0$

29. $h = 8,\ r = 10$

$V = \pi r^2 h$

$= (3.14)(10)^2 8 = 2{,}512\ in^3$

31. $V = 235.5,\ r = 15$

$V = \pi r^2 h$

$h = \frac{\pi r^2}{V} = \frac{3.14(15)^2}{235.5} = 3$ ft.

33. $V = 864,\ l = 12,\ w = 9$

(a) $V = lwh$

$h = \frac{V}{lw} = \frac{864}{12(9)} = 8$ in.

(b) $s = 2(9 \times 12) + 2(8 \times 9) + 2(8 \times 12)$

$= 552\ in^2$

35. $r = 3$

a) $V = \frac{4}{3}\pi r^3$

$= \frac{4}{3}(3.14)(3)^3 = 113.04\ cm^3$

b) $A = 4\pi r^2$

$= 4(3.14)(3)^2 = 113.04\ cm^2$

37. Triangle: $b = 16,\ h = 5$

$A = \frac{1}{2}bh = \frac{1}{2}(16)(5) = 40\ yd^2$

Quarter - circle: $r = 5$

$A = \frac{1}{4}\pi r^2 = \frac{1}{4}(3.14)(5)^2 = 19.625\ yd^2$

Total Area $= 40 + 19.625 = 59.625\ yd^2$

Cost $= (0.40)(59.625) = \$23.85$

39. Outer circle: $r = 10$

$A = \pi r^2 = 3.14(10)^2 = 314\ in^2$

Inner circle: $r = 3$

$A = \pi r^2 = 3.14(3)^2 = 28.26\ in^2$

a) Plate Area $= 314 - 28.26 = 285.74\ in^2$

b) Cost $= (7.50)(285.74) = \$2{,}143.05$

41. $l = 18,\ w = 4,\ h = 0.5$

a) $V = lwh = 18(4)(0.5) = 36\ yd^3$

b) Cost $= (3.50)(36) = \$126$

43. a) $r = 3$ in

$A = 4\pi r^2 = 4\pi(3)^2 = 36\pi\ in^2$

b) $r = 2,\ h = 0.5$

$A = 2\pi rh + 2\pi r^2$

$= 2\pi(2)(0.5) + 2\pi(2)^2$

$= 10\pi\ in^2$

The sphere has more area

Cumulative Review Problems

45. $2(x-6)-3[4-x(x+2)]$

$= 2x - 12 - 3(4 - x^2 - 2x)$

$= 2x - 12 - 12 + 3x^2 + 6x$

$= 3x^2 + 8x - 24$

47. $-\{3 - 2[x - 3(x+1)]\}$

$= -\{3 - 2(x - 3x - 3)\}$

$= -\{3 - 2x + 6x + 6\}$

$= -\{9 + 4x\}$

$= -4x - 9$

Putting Your Skills to Work

1. $A = (8 + 6 + 8 + 6)(6) + 2\left(\dfrac{1}{2}\right)(6)(3)$

$= 186 \text{ ft}^2$

3. Labor: $\dfrac{1}{2} + \dfrac{1}{2} + \dfrac{1}{2} = \dfrac{3}{2}$ days

2 hours

$\text{Cost} = 100\left(\dfrac{3}{2}\right) + 2(15) = \$180$

5. If the 2 hours of labor for clean up

is considered $\dfrac{1}{4}$ day, the labor

would be $\dfrac{7}{4}$ days at a cost of \$175.

Estimate $= 0.20(257) + 257 = \$308.40$

Exercise 3.6

1. $x > 67,000$

3. $n \le 120$

5. $P < 34$

7. $E \ge 3500$

9. $T \le 350$

11. $x =$ length of 3rd side

$180 + 135 + x \le 539$

$x \le 224$

Length of 3rd side must be not
more than 224 ft.

13. $x =$ score on final test

$\dfrac{85 + 77 + 68 + x}{4} \ge 80$

$230 + x \ge 320$

$x \ge 90$

She must make 90 or more.

15. $x =$ no. clients

$25x + 230 \le 655$

$25x \le 425$

$x \le 17$

She can take up to 17 clients to lunch.

17. $l =$ length

$12l \ge 204$

$l \ge 17$

Length must be at least 17 ft.

19. $x =$ no. of cars

$130x + 850 \ge 2800$

$130x \ge 1950$

$x \ge 15$

He must sell at least 15 cars

21. $x =$ value of products sold

$0.03x + 900 > 2700$

$0.03x > 1800$

$x > 60,000$

She must sell more than \$60,000 worth.

23. $\dfrac{5}{9}(F - 32) < 110$

$F - 32 < 198$

$F < 230$

Temperature less than 230° F.

25. $x =$ number of chairs

$$28x + 55(9) \le 850$$

$$28x \le 355$$

$$x \le 12.7$$

He can buy not more than 12 chairs.

27. $18n > 5000 + 7n$

$$11n > 5000$$

$$n > 454.5$$

Need to manufacture and sell more than 454 discs.

Cumulative Review Problems

29. $10 - 3x > 14 - 2x$

$$-x > 4$$

$$x < -4$$

31. $30 - 2(x + 1) \le 4x$

$$30 - 2x - 2 \le 4x$$

$$-6x \le -28$$

$$x \ge \frac{14}{3}$$

$$x \ge 4\frac{2}{3}$$

Chapter 3 - Review Problems

1. $x + 19$

3. $x - 56$

5. $2x + 7$

7. $5x - \dfrac{x}{3}$

9. $h =$ speed of helicopter

$h + 370 =$ speed of jet

11. $w =$ width

$3w + 5 =$ length

13. $b =$ the number of degrees in angle B

$2b =$ the number of degrees in angle A

$b - 17 =$ the number of degrees in angle C

15. $x =$ the number

$$2x + 5 = 29$$

$$2x = 24$$

$$x = 12$$

Number is 12

17. $x =$ original number

$$\frac{x}{4} - 2 = 9$$

$$\frac{x}{4} = 11$$

$$x = 44$$

Original number is 44

19. $x =$ David's age

$2x =$ Jon's age

$$2x = 32$$

$$x = 16$$

David is 16 years old

21. $t_1 =$ time for first car

$t_2 =$ time for other car

$$330 = 50t_1$$

$$6.6 = t_1$$

$$330 = 55t_2$$

$$6 = t_2$$

The first car took 6.6 hours.

The other car took 6 hours.

23. $w =$ width

$2w + 3 =$ length

$$P = 2L + 2W$$

$$2(2w + 3) + 2w = 66$$

$$4w + 6 + 2w = 66$$

$$6w = 60$$

$$w = 10, \text{ and}$$

$$2w + 3 = 23$$

Length is 23 m and width is 10 m.

25. x = length of 1st side,
   $2x - 1$ = length of 2nd side,
   $x + 9$ = length of 3rd side
   $x + 2x - 1 + x + 9 = 40$
   $$4x = 32$$
   $$x = 8, \text{ also}$$
   $$2x - 1 = 15, \text{ and}$$
   $$x + 9 = 17$$
   The lengths are 8 yd, 15 yd, 17 yd

27. x = length of one piece
   $\dfrac{3}{5}x$ = length of other
   $x + \dfrac{3}{5}x = 50$
   $5x + 3x = 250$
   $$8x = 250$$
   $$x = 31.25, \text{ and}$$
   $\dfrac{3}{5}x = 18.75$
   The lengths are 31.25 yd, 18.75 yd

29. x = number of miles driven
   $0.25x + 39(3) = 187$
   $0.25x = 70$
   $$x = 280$$
   He drove 280 miles

31. x = original price
   $0.18x = 36$
   $$x = 200$$
   Original price was $200

33. x = amount invested at 12%,
   $9000 - x$ = amount at 8%,
   $0.12x + 0.08(9000 - x) = 1000$
   $12x + 8(9000 - x) = 100,000$
   $12x + 72,000 - 8x = 100,000$
   $$4x = 28,000$$
   $$x = 7000, \text{ and}$$
   $$9000 - x = 2000$$
   They invested $7000 at 12% and
   $2000 at 8%.

35. x = number of quarters
   $x - 3$ = number of dimes
   $2x$ = number of nickels
   $0.05(2x) + 0.10(x - 3) + 0.25x = 3.75$
   $5(2x) + 10(x - 3) + 25x = 375$
   $10x + 10x - 30 + 25x = 375$
   $$45x = 405$$
   $$x = 9, \text{ also}$$
   $$x - 3 = 6, \text{ and}$$
   $$2x = 18$$
   She has 18 nickels, 6 dimes, 9 quarters

37. $C = 25.12$
   $C = 2\pi r = \pi d$
   $d = \dfrac{C}{\pi} = \dfrac{25.12}{3.14} = 8\ m$

39. x = measure of third angle
   $62 + 47 + x = 180$
   $$x = 71^0$$
   The angle is 71^0

41. Smallest side = 8
   Second side = $2(8) = 16$
   Third side = $2(16) = 32$
   $P = 8 + 16 + 32 = 56$ in.

43. $l = 12, \ w = 8, \ h = 20$
   $V = lwh = 12(8)(20) = 1920 \text{ ft}^3$

45. x = length of 3rd side
   P = sum of sides
   $17 + 17 + x = 46$
   $$x = 12$$
   The 3rd side is 12 in.

47. $r = 5, h = 14$
   $A = 2\pi rh + 2\pi r^2$
   $= 2(3.14)(5)(14) + 2(3.14)(5)^2$
   $= 439.60 + 157.00$
   $= 596.60\ m^2$
   Cost $= 40(596.60) = \$23,864.00$

49. x = score on final

$$\frac{92+85+78+86+x}{5} \geq 85$$

$$341+x \geq 425$$

$$x \geq 84$$

He must score at least 84.

51. $x =$ miles driven

$$0.15x+2(35) \leq 100$$

$$15x+7000 \leq 10,000$$

$$15x \leq 3000$$

$$x \leq 200$$

Distance must be no more than 200 miles.

53. x = amount for ties

$$3(17.95)+x < 70$$

$$x < 16.15$$

He must spend less than $16.15 for ties.

55. x = length of commercials,
$4x+4$ = length of entertainment

$$x+4x+4 = 30$$

$$5x = 26$$

$$x = 5.2, \text{ and}$$

$$4x+4 = 24.8$$

Entertainment was 24.8 min.

Putting Your Skills To Work

1. Boat: $r = 2$

$d = rt = 2t$

Helicopter: $r = 150$

$d = rt = 150t$

$$d_{\text{total}} = d_B + d_H$$

$$76 = 2t+150t$$

$$76 = 152t$$

$$0.5 = t$$

It will reach the boat in 0.5 hours, or 30 minutes.

Test - Chapter 3

1. x = original number

$$2x-11 = 59$$

$$2x = 70$$

$$x = 35$$

Original number is 35.

3. x = original number

$$3x+6 = 2(x-3)$$

$$3x+6 = 2x-6$$

$$x = -12$$

Original number is -12.

5. w = width

$2w+7$ = length

$P = 2L+2W$

$$2(2w+7)+2w = 134$$

$$4w+14+2w = 134$$

$$6w = 120$$

$$w = 20, \text{ and}$$

$$2w+7 = 47$$

Rectangle is 47 m by 20 m.

7. x = number of months

$$116x+200 = 1940$$

$$116x = 1740$$

$$x = 15$$

He can rent 15 months.

9. x = amount invested at 14%,
$4000-x$ = amount at 11%

$$0.14x+0.11(4000-x) = 482$$

$$14x+11(4000-x) = 48,200$$

$$14x+44,000-11x = 48,200$$

$$3x = 4200$$

$$x = 1400, \text{ and}$$

$$4000-x = 2600$$

He invested $1400 at 14% and $2600 at 11%.

11. $r = 22$

$C = 2\pi r = 2(3.14)(22) = 138.16$ in.

13. $r = 10$

$V = \frac{4}{3}\pi r^3$

$= \frac{4}{3}(3.14)(10)^3$

$= 4187$ in^3

15. $A = 12(1.5) + 9(2) + \frac{1}{2}(2)(1.5)$

$= 18 + 18 + 1.5 = 37.5$ yd^2

Cost $= 12(37.5) = \$450$

17. $x =$ sales over 10,000

$0.05x + 15,000 > 20,000$

$0.05x > 5000$

$x > 100,000$

She must sell more than \$100,000

$+ 10,000 = 110,000.$

Cumulative Test - Chapters 0 - 3

1.
$$
\begin{array}{r}
3.69 \\
2.4\overline{)8.856} \\
\underline{72} \\
165 \\
\underline{144} \\
216 \\
\underline{216}
\end{array}
$$

3. $3(2x - 3y + 6) + 2(-3x - 7y)$

$= 6x - 9y + 18 - 6x - 14y$

$= -23y + 18$

5.
$$H = \frac{1}{2}(3a + 5b)$$

$$2H = 2\left(\frac{1}{2}\right)(3a + 5b)$$

$$2H = 3a + 5b$$

$$2H + (-3a) = 3a + (-3a) + 5b$$

$$2H - 3a = 5b$$

$$\frac{2H - 3a}{5} = \frac{5b}{5}$$

$$\frac{2H - 3a}{5} = b$$

7. $x =$ original number

$3x - 17 = 103$

$3x = 120$

$x = 40$

Original number is 40.

9. $w =$ width

$3w + 11 =$ length

$P = 2L + 2W$

$2(3w + 11) + 2w = 78$

$6w + 22 + 2w = 78$

$8w = 56$

$w = 7$, and

$3w + 11 = 32$

Dimensions are 32 cm by 7 cm.

11. $x =$ amount invested at 15%,

$7000 - x =$ amount invested at 7%

$0.15x + 0.07(7000 - x) = 730$

$15x + 7(7000 - x) = 73,000$

$15x + 49,000 - 7x = 73,000$

$8x = 24,000$

$x = 3000$, and

$7000 - x = 4000$

He invested \$3000 at 15% and
\$4000 at 7%.

13. $b = 25$, $h = 13$

$A = \frac{1}{2}bh = \frac{1}{2}(25)(13) = 162.5 \ m^2$

Cost $= 4.50(162.5) = \$731.25$

Practice Quiz: Sections 3.1 - 3.3

In questions 1 - 3, write an algebraic
expression. Use the variable x to
represent the unknown number.

1. Two more than three times a number.

2. Five less than one - fifth of a number.

3. One - half of three more than twice
   a number.

4. A number is increased by six and then divided by three. The result is 6. Find the original number.

5. The length of a rectangle is 4 centimeters longer than twice the width. The perimeter is 38 centimeters. Find the dimensions of the rectangle.

Practice Quiz: Sections 3.4 - 3.6

1. Emilio invested $3000 in two different certificates. One yielded 6% and the other 8%. The total interest from both investments was $212. How much was invested at each rate?

2. Louise's salary was increased by 7% over last year's salary. This year's salary is $24,610. What was last year's salary?

3. Find the area of a parallelogram which has a base of 7 in and a height of 5 in.

4. Find the volume of a right circular cylinder which has a height of 8 centimeters and whose base has a radius of 4 centimeters. Use 3.14 for π.

5. Erin gets a 2.5% commission on sales in addition to a base salary of $15,000. How much does she need to sell if she wants to make more than $24,000?

Answers to Practice Quiz

Sections 3.1 - 3.3

1. $3x + 2$

2. $\frac{x}{5} - 5$

3. $\frac{1}{2}(2x + 3)$

4. 12

5. 14 cm by 5 cm

Answers to Practice Quiz

Sections 3.4 - 3.6

1. $1400 at 6% and $1600 at 8%

2. $23,000

3. 35 in^2

4. 401.92 cm^2

5. More than $360,000

Pretest Chapter 4

1. $(3^6)(3^{10}) = 3^{6+10} = 3^{16}$

3. $(-4ab^2)(2a^3b)(3ab) = -24a^5b^4$

5. $\dfrac{12xy^2}{-6x^3y^4} = -\dfrac{2}{x^{3-1}y^{4-2}} = -\dfrac{2}{x^2y^2}$

7. $(x^5)^{10} = x^{5(10)} = x^{50}$

9. $\left(\dfrac{3ab^2}{c^3}\right)^2 = \dfrac{3^2a^2b^{2(2)}}{c^{3(2)}} = \dfrac{9a^2b^4}{c^6}$

11. $\dfrac{-2a^{-3}}{b^{-2}c^4} = \dfrac{-2b^2}{a^3c^4}$

13. $0.000638 = 6.38 \times 10^{-4}$

15. $(3x^2 - 6x - 9) + (-5x^2 + 13x - 20)$
$= (3-5)x^2 + (-6+13)x - 9 - 20$
$= -2x^2 + 7x - 29$

17. $-3x(2 - 7x) = -6x + 21x^2$

19. $(x^3 - 3xy + y^2)(5xy^2)$
$= 5x^4y^2 - 15x^2y^3 + 5xy^4$

21. $(3x - 2y)(4x + 3y)$
$= 12x^2 + 9xy - 8xy - 6y^2$
$= 12x^2 + xy - 6y^2$

23. $(8x - 11y)(8x + 11y) = (8x)^2 - (11y)^2$
$= 64x^2 - 121y^2$

25. $(5ab - 6)^2 = (5ab)^2 + 2(5ab)(-6) + (-6)^2$
$= 25a^2b^2 - 60ab + 36$

27. $(2x - 3)(2x + 3)(4x^2 + 9)$
$= (4x^2 - 9)(4x^2 + 9)$
$- 16x^4 - 81$

29. $$3x - 2 \overline{)\,5x^2 - 6x + 2 \atop 15x^3 - 28x^2 + 18x - 7}$$
$$\underline{15x^3 - 10x^2}$$
$$-18x^2 + 18x$$
$$\underline{-18x^2 + 12x}$$
$$6x - 7$$
$$\underline{6x - 4}$$
$$-3$$

$(15x^3 - 28x^2 + 18x - 7) \div (3x - 2)$

$= 5x^2 - 6x + 2 - \dfrac{3}{3x - 2}$

Exercises 4.1

1. Answers may vary. See text explanation.

3. $\dfrac{2^2}{2^3} \overset{?}{=} \dfrac{1}{2^{3-2}}$

$\dfrac{4}{8} \overset{?}{=} \dfrac{1}{2}$

$\dfrac{1}{2} = \dfrac{1}{2}$

5. $6x^{11}y$: Coefficient is 6, Bases are x, y,
Exponents are 11, 1.

7. $3 \cdot x \cdot x \cdot x \cdot y \cdot y = 3x^3y^2$

9. $(-3)(a)(a)(b)(c)(b)(c)(c) = -3a^2b^2c^3$

11. $(3^8)(3^7) = 3^{8+7} = 3^{15}$

13. $(5^{10})(5^{16}) = 5^{10+16} = 5^{26}$

15. $(3^5)(8^2) = 3^5 \cdot 8^2$

17. $-12x^2 \cdot 5x = -60x^{2+1} = -60x^3$

19. $(-13x)(-2x^2) = 26x^{1+2} = 26x^3$

21. $(4x^8)(-3x^3) = -12x^{8+3} = -12x^{11}$

23. $(5xy^2)(-2x^3y) = -10x^{1+3}y^{2+1} = -10x^4y^3$

25. $\left(\dfrac{2}{3}x^3y\right)\left(\dfrac{5}{7}xy^4\right) = \dfrac{10}{21}x^{3+1}y^{1+4} = \dfrac{10}{21}x^4y^5$

27. $(2.3x^4w)(-3.5xy^4) = -8.05x^{4+1}y^4w$
$$= -8.05x^5y^4w$$

29. $(8a)(2a^3b)(0) = 0$

31. $(-4x)(2x^2y)(3x) = -24x^{1+2+1}y$
$$= -24x^4y$$

33. $(-6xy^3)(-12x^5y^8) = 72x^{1+5}y^{3+8}$
$$= 72x^6y^{11}$$

35. $(14a^5)(-2b^6) = -28a^5b^6$

37. $(-x^2y)(3x) = -3x^{2+1}y = -3x^3y$

39. $(3ab^3)(-2a^5b)(-2a^2b^2)$
$$= 12a^{1+5+2}b^{3+1+2}$$
$$= 12a^8b^6$$

41. $(4x^3y)(0)(-12x^4y^2) = 0$

43. $(6w^5z^6)(-4xy) = -24w^5xyz^6$

45. $\dfrac{x^8}{x^{10}} = \dfrac{1}{x^{10-8}} = \dfrac{1}{x^2}$

47. $\dfrac{y^{12}}{y^5} = y^{12-5} = y^7$

49. $\dfrac{3^{45}}{3^{17}} = 3^{45-17} = 3^{28}$

51. $\dfrac{13^{20}}{13^{30}} = \dfrac{1}{13^{30-20}} = \dfrac{1}{13^{10}}$

53. $\dfrac{2^{20}}{2^7} = 2^{20-7} = 2^{13}$

55. $\dfrac{a^{13}}{4a^5} = \dfrac{a^{13-5}}{4} = \dfrac{a^8}{4}$

57. $\dfrac{x^7}{y^9} = \dfrac{x^7}{y^9}$

59. $\dfrac{-10a^5b^2}{-20a^5b^3} = \dfrac{a^{5-5}}{2b^{3-2}} = \dfrac{1}{2b}$

61. $\dfrac{12x^5y^6}{-3xy} = -4x^{5-1}y^{6-1} = -4x^4y^5$

63. $\dfrac{12ab}{-24a^3b^2} = \dfrac{1}{-2a^{3-1}b^{2-1}} = \dfrac{-1}{2a^2b}$

65. $\dfrac{-13x^5y^6}{13x^5y^6} = -1x^{5-5}y^{6-6} = -1$

67. $\dfrac{-27x^5y^3z}{9x^3y^6z^2} = \dfrac{-3x^{5-3}}{y^{6-3}z^{2-1}} = \dfrac{-3x^2}{y^3z}$

69. $\dfrac{2.3x^8y^{10}}{23x^{12}y^5} = \dfrac{y^{10-5}}{10x^{12-8}} = \dfrac{y^5}{10x^4}$

71. $\dfrac{30x^5y^4}{5x^3y^4} = 6x^{5-3}y^{4-4} = 6x^2$

73. $\dfrac{8^0x^2y^3}{16x^5y} = \dfrac{y^{3-1}}{16x^{5-2}} = \dfrac{y^2}{16x^3}$

75. $\dfrac{18a^6b^3c^0}{24a^5b^3} = \dfrac{3}{4}a^{6-5}b^{3-3}c^0 = \dfrac{3}{4}a$

77. $\dfrac{25x^6}{35y^8} = \dfrac{5x^6}{7y^8}$

79. $\dfrac{-32x^6y^3z}{64x^5y^2} = -\dfrac{1}{2}x^{6-5}y^{3-2}z$
$$= -\dfrac{1}{2}xyz$$

81. $\dfrac{(3x^2)(2x^3)}{3x^4} = \dfrac{6x^5}{3x^4} = 2x$

83. $\dfrac{(9a^2b)(2a^3b^6)}{-27a^8b^7} = \dfrac{18a^5b^7}{-27a^8b^7} = -\dfrac{2}{3a^3}$

85. $\dfrac{63a^5b^6}{-9a^4b} = -7ab^5$

87. Let $y = 2$: $\dfrac{y^8}{y^4} = \dfrac{2^8}{2^4} = \dfrac{256}{16} = 16$
$$y^2 = 2^2 = 4$$
$$\dfrac{y^8}{y^4} \neq y^2$$

89. $(a^{2b})(a^{3b}) = a^{2b+3b} = a^{5b}$

64

91. $\dfrac{\left(c^{3y}\right)\left(c^4\right)}{c^{2y+2}} = c^{3y+4-2y-2} = c^{y+2}$

93. $\left(w^5\right)^8 = w^{5\cdot 8} = w^{40}$

95. $\left(a^3b^2\right)^5 = a^{3\cdot 5}b^{2\cdot 5} = a^{15}b^{10}$

97. $\left(a^3b^2c\right)^8 = a^{3\cdot 8}b^{2\cdot 8}c^{18} = a^{24}b^{16}c^8$

99. $\left(3^2 xy^2\right)^4 = 3^{2\cdot 4}x^{1\cdot 4}y^{2\cdot 4} = 3^8 x^4 y^8$

101. $\left(-2a^5\right)^4 = (-2)^4 a^{5\cdot 4} = 16a^{20}$

103. $\left(\dfrac{12x}{y^2}\right)^5 = \dfrac{12^5 x^{1\cdot 5}}{y^{2\cdot 5}} = \dfrac{12^5 x^5}{y^{10}}$

105. $\left(\dfrac{6x}{5y^3}\right)^2 = \dfrac{6^2 x^{1\cdot 2}}{5^2 y^{3\cdot 2}} = \dfrac{36x^2}{25y^6}$

107. $\left(-2a^5b^2c\right)^5 = (-2)^5 a^{5\cdot 5}b^{2\cdot 5}c^{1\cdot 5}$
$$= -32a^{25}b^{10}c^5$$

109. $\left(-4xy^0z^4\right)^3 = (-4)^3 x^3 y^{0\cdot 3}z^{4\cdot 3}$
$$= -64x^3 z^{12}$$

111. $\dfrac{\left(4a^2b\right)^2}{\left(4ab^2\right)^3} = \dfrac{4^2 a^4 b^2}{4^3 a^3 b^6} = \dfrac{a}{4b^4}$

113. $\left(-2a^2b^3\right)^3\left(ab^2\right) = \left[(-2)^3 a^6 b^9\right]\left(ab^2\right)$
$$= -8a^7 b^{11}$$

115. $\left(\dfrac{4x^0y^4}{3z^3}\right)^2 = \dfrac{4^2 x^0 y^8}{3^2 z^6} = \dfrac{16y^8}{9z^6}$

117. $\left(\dfrac{y^6}{2}\right)^3 = \dfrac{y^{18}}{2^3} = \dfrac{y^{18}}{8}$

119. $\left(\dfrac{a^3b}{c^5d}\right)^5 = \dfrac{a^{15}b^5}{c^{25}d^5}$

121. $\left(4x^3y\right)^2\left(x^3y\right) = \left(4^2 x^6 y^2\right)\left(x^3y\right)$
$$= 16x^9 y^3$$

123. $\left(\dfrac{3a^3b}{c^2}\right)^4 = \dfrac{3^4 a^{12}b^4}{c^8} = \dfrac{81a^{12}b^4}{c^8}$

125. $16x^{20}y^{16}z^{28} = \left(\pm 2x^5 y^4 z^7\right)^4$
Therefore, $\pm 2x^5 y^4 z^7$

127. Let $y = 3$; $(y+2)^2 = (3+2)^2 = 25$
$$y^2 + 2^2 = 3^2 + 2^2 = 13$$
$$(y+2)^2 \neq y^2 + 2^2$$

129. $4^2 \cdot 8^3 \cdot 16 = \left(2^2\right)^2\left(2^3\right)^3\left(2^4\right)$
$$= 2^4 \cdot 2^9 \cdot 2^4 = 2^{17}$$

Cumulative Review Problems

131. $-3-8 = -3 + (-8) = -11$

133. $\left(\dfrac{2}{3}\right)\left(\dfrac{-21}{8}\right) = \left(\dfrac{2}{3}\right)\left(\dfrac{-3\cdot 7}{4\cdot 2}\right) = -\dfrac{7}{4}$

Exercises 4.2

1. $3x^{-2} = \dfrac{3}{x^2}$

3. $(2xy)^{-1} = \dfrac{1}{2xy}$

5. $\dfrac{3xy^{-2}}{z^{-3}} = \dfrac{3xz^3}{y^2}$

7. $\dfrac{(3x)^{-2}}{(3x)^{-3}} = \dfrac{(3x)^3}{(3x)^2} = 3x$

9. $x^6 y^{-2} z^{-3} w^{-10} = \dfrac{x^6}{y^2 z^3 w^{10}}$

11. $\left(4^{-3}\right)\left(2^0\right) = \dfrac{1}{4^3} = \dfrac{1}{64}$

13. $\left(\dfrac{3xy^2}{z^4}\right)^{-2} = \left(\dfrac{z^4}{3xy^2}\right)^2 = \dfrac{z^8}{9x^2 y^4}$

15. $\dfrac{x^{-2}y^{-3}}{x^4 y^{-2}} = \dfrac{y^2}{x^4 x^2 y^3} = \dfrac{1}{x^6 y}$

17. $\left(-2x^3 y^{-2}\right)^{-3} = \left(\dfrac{-2x^3}{y^2}\right)^{-3}$

$\quad = \left(\dfrac{y^2}{-2x^3}\right)^3$

$\quad = -\dfrac{y^6}{8x^9}$

19. $123{,}780 = 1.2378 \times 10^5$

21. $0.000742 = 7.42 \times 10^{-4}$

23. $7{,}652{,}000{,}000 = 7.652 \times 10^9$

25. $5.63 \times 10^4 = 56{,}300$

27. $3.3 \times 10^{-5} = 0.000033$

29. $9.83 \times 10^5 = 983{,}000$

31. $3.00 \times 10^8_{m/s} = 300{,}000{,}000 \ m/s$

33. $0.00000000000000000000000000000911 \ kg$
$\quad = 9.11 \times 10^{-31} \ kg$

35. $\dfrac{(5{,}000{,}000)(16{,}000)}{8{,}000{,}000{,}000} = \dfrac{\left(5 \times 10^6\right)\left(1.6 \times 10^4\right)}{8 \times 10^9}$

$\quad = \dfrac{8 \times 10^{10}}{8 \times 10^9}$

$\quad = 1.0 \times 10^1$

37. $(0.0002)^5 = \left(2 \times 10^{-4}\right)^5$

$\quad = 2^5 \times 10^{-20}$

$\quad = 32 \times 10^{-20}$

$\quad = 3.2 \times 10^{-19}$

39. $(150{,}000{,}000)(0.00005)(0.002)(30{,}000)$

$\quad = \left(1.5 \times 10^8\right)\left(5 \times 10^{-5}\right)\left(2 \times 10^{-3}\right)\left(3 \times 10^4\right)$

$\quad = 45 \times 10^4$

$\quad = 4.5 \times 10^5$

41. $\dfrac{3.502 \times 10^{12}}{2.49 \times 10^8} = 1.406426 \times 10^4$

$\quad = 1.41 \times 10^4 \ / \ \text{person}$

43. $d = (276)\left(3.09 \times 10^{13}\right)$

$\quad = \left(2.76 \times 10^2\right)\left(3.09 \times 10^{13}\right) = 8.528 \times 10^{15}$

$\quad r = 45{,}000 = 4.5 \times 10^4$

$\quad d = rt$

$\quad t = \dfrac{d}{r} = \dfrac{8.528 \times 10^{15}}{4.5 \times 10^4} = 1.90 \times 10^{11} \ \text{hours}$

45. $\quad m = 9.11 \times 10^{-28}$

$\quad \text{number} = 125{,}000{,}000 = 1.25 \times 10^8$

$\quad m_{\text{total}} = \left(1.25 \times 10^8\right)\left(9.11 \times 10^{-28}\right)$

$\quad\quad = 11.3875 \times 10^{-20}$

$\quad\quad = 1.13875 \times 10^{-19} \ g$

Cumulative Review Problems

47. $-2.7 - (-1.9) = -2.7 + 1.9 = -0.8$

49. $\dfrac{-3}{4} + \dfrac{5}{7} = \dfrac{-21}{28} + \dfrac{20}{28} = \dfrac{-21 + 20}{28} = -\dfrac{1}{28}$

Putting Your Skills to Work

1. $M_1 = 5.97 \times 10^{24}$

$\quad M_2 = 7.35 \times 10^{22}$

$\quad d = 3.84 \times 10^8$

$\quad F = \dfrac{6.67 \times 10^{-11}(M_1)(M_2)}{d^2}$

$\quad = \dfrac{6.67 \times 10^{-11}\left(5.97 \times 10^{24}\right)\left(7.35 \times 10^{22}\right)}{\left(3.84 \times 10^8\right)^2}$

$\quad = \dfrac{292.68 \times 10^{35}}{14.746 \times 10^{16}}$

$\quad = 19.848 \times 10^{19}$

$\quad = 1.98 \times 10^{20} \ N$

3. $M_1 = 1.90 \times 10^{27}$

$M_2 = 4.85 \times 10^{19}$

$d = 6.71 \times 10^8$

$$F = \frac{6.67 \times 10^{-11}(M_1)(M_2)}{d^2}$$

$$= \frac{6.67 \times 10^{-11}(1.90 \times 10^{27})(4.85 \times 10^{19})}{(6.71 \times 10^8)^2}$$

$$= \frac{61.464 \times 10^{35}}{45.024 \times 10^{16}}$$

$$= 1.37 \times 10^{19} \, N$$

5. The force in problem 1 is greater because the product of the masses is greater and the distance is less.

7. Doubling each mass would make the product, and therefore the force, four times as large.

Exercises 4.3

1. $(7x+3)+(-5x-20) = (7-5)x+3-20$

$$= 2x - 17$$

3. $(x^2+5x+6)+(-2x^2-6x+8)$

$$= (1-2)x^2+(5-6)x+6+8$$

$$= -x^2 - x + 14$$

5. $(x^3-3x^2+7)+(x^2+5x-8)$

$$= x^3+(-3+1)x^2+5x+7-8$$

$$= x^3 - 2x^2 + 5x - 1$$

7. $\left(\dfrac{1}{2}x^2+\dfrac{1}{3}x-4\right)+\left(\dfrac{1}{3}x^2+\dfrac{1}{6}x-5\right)$

$$= \left(\dfrac{1}{2}+\dfrac{1}{3}\right)x^2+\left(\dfrac{1}{3}+\dfrac{1}{6}\right)x-4-5$$

$$= \dfrac{5}{6}x^2+\dfrac{1}{2}x-9$$

9. $(3.4x^3-5.6x^2-7.1x+3.4)$

$+(-1.7x^3+2.2x^2-6.1x-8.8)$

$$= (3.4-1.7)x^3+(-5.6+2.2)x^2$$

$$+(-7.1-6.1)x+3.4-8.8$$

$$= 1.7x^3 - 3.4x^2 - 13.2x - 5.4$$

11. $(-8x+2)-(3x+4) = (-8x+2)+(-3x-4)$

$$= (-8-3)x+2-4$$

$$= -11x - 2$$

13. $(5x^3-7x^2)-(-12x^3-8x^2)$

$$= (5x^3-7x^2)+(12x^3+8x^2)$$

$$= (5+12)x^3+(-7+8)x^2$$

$$= 17x^3 + x^2$$

15. $\left(\dfrac{1}{2}x^2+3x-\dfrac{1}{5}\right)-\left(\dfrac{2}{3}x^2-4x+\dfrac{1}{10}\right)$

$$= \left(\dfrac{1}{2}x^2+3x-\dfrac{1}{5}\right)+\left(-\dfrac{2}{3}x^2+4x-\dfrac{1}{10}\right)$$

$$= \left(\dfrac{1}{2}-\dfrac{2}{3}\right)x^2+(3+4)x-\dfrac{1}{5}-\dfrac{1}{10}$$

$$= -\dfrac{1}{6}x^2+7x-\dfrac{3}{10}$$

17. $(-3x^2+5x)-(2x^3-3x^2+10)$

$$= (-3x^2+5x)+(-2x^3+3x^2-10)$$

$$= -2x^3+(-3+3)x^2+5x-10$$

$$= -2x^3 + 5x - 10$$

19. $(0.5x^4-0.7x^2+8.3)-(5.2x^4+1.6x+7.9)$

$$= (0.5x^4-0.7x^2+8.3)+(-5.2x^4-1.6x-7.9)$$

$$= (0.5-5.2)x^4-0.7x^2-1.6x+8.3-7.9$$

$$= -4.7x^4 - 0.7x^2 - 1.6x + 0.4$$

21. $(7x-8)-(5x-6)+(8x+12)$

$$= (7x-8)+(-5x+6)+(8x+12)$$

$$= (7-5+8)x-8+6+12$$

$$= 10x + 10$$

23. $\left(5x^2y - 6xy^2 + 2\right) + \left(-8x^2y + 12xy^2 - 6\right)$

$= (5-8)x^2y + (-6+12)xy^2 + 2 - 6$

$= -3x^2y + 6xy^2 - 4$

25. $\left(-7a^4 + 3a^2 - 2\right) - \left(4a^4 + 5a^3 + 3a^2\right)$

$= \left(-7a^4 + 3a^2 - 2\right) + \left(-4a^4 - 5a^3 - 3a^2\right)$

$= (-7-4)a^4 - 5a^3 + (3-3)a^2 - 2$

$= -11a^4 - 5a^3 - 2$

27. $\left(-2x^3 - 3x^2y + 4xy^2 - 7y^3\right)$

$+ \left(x^3 + x^2y - 9xy^2 + 2y^3\right)$

$= (-2+1)x^3 + (-3+1)x^2y$

$+ (4-9)xy^2 + (-7+2)y^3$

$= -x^3 - 2x^2y - 5xy^2 - 5y^3$

29. $\left(-63x^2 + 34xy + y^2\right) + \left(5x^2 - 7xy - 9y^2\right)$

$= (-63+5)x^2 + (34-7)xy + (1-9)y^2$

$= -58x^2 + 27xy - 8y^2$

31. $\left(12x^3 + 34x^2 - 6\right) - \left(5x^3 - 7x^2 - 28\right)$

$= \left(12x^3 + 34x^2 - 6\right) + \left(-5x^3 + 7x^2 + 28\right)$

$= (12-5)x^3 + (34+7)x^2 - 6 + 28$

$= 7x^3 + 41x^2 + 22$

Cumulative Review Problems

33. $3y - 8x = 2$

$-8x = -3y + 2$

$x = \dfrac{-3y+2}{-8}$

$x = \dfrac{3y-2}{8}$

35. $A = \dfrac{1}{2}h(b+c)$

$2A = h(b+c)$

$\dfrac{2A}{h} = b + c$

$\dfrac{2A}{h} - c = b$ or $b = \dfrac{2A - hc}{h}$

1. $3x\left(5x^2 - x\right) = 15x^3 - 3x^2$

3. $-4x\left(x^2 - 3x - 6\right) = -4x^3 + 12x^2 + 24x$

5. $2x\left(x^3 - 3x^2 + x - 6\right) = 2x^4 - 6x^3 + 2x^2 - 12x$

7. $-5xy\left(x^2y^2 - 3xy + 5\right)$

$= -5x^3y^3 + 15x^2y^2 - 25xy$

9. $3x^3\left(x^4 + 2x^2 - 3x + 7\right)$

$= 3x^7 + 6x^5 - 9x^4 + 21x^3$

11. $2ab^2\left(3 - 2ab - 5b^3\right)$

$= 6ab^2 - 4a^2b^3 - 10ab^5$

13. $\left(5x^3 - 2x^2 + 6x\right)\left(-3xy^2\right)$

$= -15x^4y^2 + 6x^3y^2 - 18x^2y^2$

15. $\left(2b^2 + 3b - 4\right)\left(2b^2\right) = 4b^4 + 6b^3 - 8b^2$

17. $\left(x^3 - 3x^2 + 5x - 2\right)(3x) = 3x^4 - 9x^3 + 15x^2 - 6x$

19. $\left(x^2y^2 - 6xy + 8\right)(-2xy)$

$= -2x^3y^3 + 12x^2y^2 - 16xy$

21. $\left(-7x^3 + 3x^2 + 2x - 1\right)\left(4x^2y\right)$

$= -28x^5y + 12x^4y + 8x^3y - 4x^2y$

23. $\left(5a^3 + a^2 - 7\right)\left(-3ab^2\right)$

$= -15a^4b^2 - 3a^3b^2 + 21ab^2$

25. $6x^3\left(2x^4 - x^2 + 3x + 9\right)$

$= 12x^7 - 6x^5 + 18x^4 + 54x^3$

27. $-2x^4\left(9x^3 + 2x - 4\right) = -18x^7 - 4x^5 + 8x^4$

29. $(x + 10)(x + 3) = x^2 + 13x + 30$

31. $(x + 6)(x + 2) = x^2 + 8x + 12$

33. $(x + 3)(x - 6) = x^2 - 3x - 18$

35. $(x-6)(x-5) = x^2 - 11x + 30$

37. $(5x+1)(-x-3) = -5x^2 - 16x - 3$

39. $(3x-5)(x+3y) = 3x^2 + 9xy - 5x - 15y$

41. $(3y+2)(4y-3) = 12y^2 - y - 6$

43. $(6y-5)(2y-1) = 12y^2 - 16y + 5$

45. $-(3x-7) = -3x + 7$
    The last term should be $+7$

47. $(7x+3)(7x-3) = 49x^2 - 9$
    It is the difference of the squares
    of the two terms.

49. $(4x-2y)(-7x-3y) = -28x^2 + 2xy + 6y^2$

51. $(3a-2b^2)(5a-6b^2) = 15a^2 - 28ab^2 + 12b^4$

53. $(2x^2-5y^2)(2x^2+5y^2) = 4x^4 - 25y^4$

55. $(8x-2)^2 = (8x-2)(8x-2)$
    $= 64x^2 - 32x + 4$

57. $(5x^2+2y^2)^2 = (5x^2+2y^2)(5x^2+2y^2)$
    $= 25x^4 + 20x^2y^2 + 4y^4$

59. $(5t-9p)(3t+7p) = 15t^2 + 8tp - 63p^2$

61. $(3ab-5d)(2ab-7d) = 6a^2b^2 - 31abd + 35d^2$

63. $(5x+8y)(6x-y) = 30x^2 + 43xy - 8y^2$

65. $(7x-2y)(5x+3z) = 35x^2 + 21xz - 10xy - 6yz$

67. $(a-8b)(4c-3b) = 4ac - 3ab - 32bc + 24b^2$

69. $\left(\frac{1}{3}x^3y^4 - \frac{1}{7}xy^6\right)\left(\frac{1}{2}x^3y^4 + \frac{1}{3}xy^6\right)$

    $= \frac{1}{6}x^6y^8 + \frac{1}{9}x^4y^{10} - \frac{1}{14}x^4y^{10} - \frac{1}{21}x^2y^{12}$

    $= \frac{1}{6}x^6y^8 + \frac{5}{126}x^4y^{10} - \frac{1}{21}x^2y^{12}$

Cumulative Review Problems

71. $3(x-6) = -2(x+4) + 6x$
    $3x - 18 = -2x - 8 + 6x$
    $-10 = x$

73. Let d = number of dimes
    $d+3 =$ number of quarters
    $0.10d + 0.25(d+3) = 3.55$
    $10d + 25(d+3) = 355$
    $10d + 25d + 75 = 355$
    $35d = 280$
    $d = 8$
    $d+3 = 11$
    She has 8 dimes and 11 quarters

75. $(1.62x + 3.53y)(1.98x - 4.82y)$
    $= 3.2076x^2 - 7.8084xy + 6.9894xy - 17.0146y^2$
    $= 3.2076x^2 - 0.819xy - 17.0146y^2$

Exercises 4.5

1. In the special case of $(a+b)(a-b)$,
   a binomial times a binomial is a
   binomial.

3. $(4x-7)^2 = 16x^2 - 56x + 49$
   The student left out the middle
   term which comes from the
   product of the two outer terms
   and the product of the two
   inner terms.

5. $(x+4)(x-4) = x^2 - (4)^2 = x^2 - 16$

7. $(x-9)(x+9) = x^2 - (9)^2 = x^2 - 81$

9. $(8x+3)(8x-3) = (8x)^2 - (3)^2 = 64x^2 - 9$

11. $(2x-7)(2x+7) = (2x)^2 - (7)^2 = 4x^2 - 49$

13. $(2x-5y)(2x+5y) = (2x)^2 - (5y)^2 = 4x^2 - 25y^2$

15. $(10p - 7q)(10p + 7q) = (10p)^2 - (7q)^2$
$$= 100p^2 - 49q^2$$

17. $(12x^2 + 7)(12x^2 - 7) = (12x^2) - (7)^2$
$$= 144x^4 - 49$$

19. $(3x^2 - 4y^3)(3x^2 + 4y^3) = (3x^2)^2 - (4y^3)^2$
$$= 9x^4 - 16y^6$$

21. $(4x - 1)^2 = 16x^2 - 2(4x) + 1 = 16x^2 - 8x + 1$

23. $(8x + 3)^2 = (8x)^2 + 2(3)(8x) + (3)^2$
$$= 64x^2 + 48x + 9$$

25. $(5x - 7)^2 = (5x)^2 + 2(-7)(5x) + (-7)^2$
$$= 25x^2 - 70x + 49$$

27. $(2x + 3y)^2 = (2x)^2 + 2(2x)(3y) + (3y)^2$
$$= 4x^2 + 12xy + 9y^2$$

29. $(8x - 5y)^2 = (8x)^2 + 2(-5y)(8x) + (-5y)^2$
$$= 64x^2 - 80xy + 25y^2$$

31. $(8 + 3x)^2 = (8)^2 + 2(8)(3x) + (3x)^2$
$$= 64 + 48x + 9x^2$$

33. $(6w + 5z)^2 = (6w)^2 + 2(6w)(5z) + (5z)^2$
$$= 36w^2 + 60wz + 25z^2$$

35. $(7x + 3y)(7x - 3y) = (7x)^2 - (3y)^2$
$$= 49x^2 - 9y^2$$

37. $(4x^2 - 7y)^2 = (4x^2)^2 + 2(4x^2)(-7y) + (-7y)^2$
$$= 16x^4 - 56x^2 y + 49y^2$$

39. $(x^2 - 2x + 1)(x - 2) = x(x^2 - 2x + 1) - 2(x^2 - 2x + 1)$
$$= x^3 - 2x^2 + x - 2x^2 + 4x - 2$$
$$= x^3 - 4x^2 + 5x - 2$$

41. $(3y + 2z)(3y^2 - yz - 2z^2)$
$$= 3y(3y^2 - yz - 2z^2) + 2z(3y^2 - yz - 2z^2)$$
$$= 9y^3 - 3y^2 z - 6yz^2 + 6y^2 z - 2yz^2 - 4z^3$$
$$= 9y^3 + 3y^2 z - 8yz^2 - 4z^3$$

43. $(4x + 1)(x^3 - 2x^2 + x - 1)$
$$= 4x(x^3 - 2x^2 + x - 1) + 1(x^3 - 2x^2 + x - 1)$$
$$= 4x^4 - 8x^3 + 4x^2 - 4x + x^3 - 2x^2 + x - 1$$
$$= 4x^4 - 7x^3 + 2x^2 - 3x - 1$$

45. $(x + 2)(x - 3)(2x + 5)$
$$= (x^2 - x - 6)(2x + 5)$$
$$= (x^2 - x - 6)(2x) + (x^2 - x - 6)(5)$$
$$= 2x^3 - 2x^2 - 12x + 5x^2 - 5x - 30$$
$$= 2x^3 + 3x^2 - 17x - 30$$

47. $(3x + 5)(x - 2)(x - 4)$
$$= (3x + 5)(x^2 - 6x + 8)$$
$$= 3x(x^2 - 6x + 8) + 5(x^2 - 6x + 8)$$
$$= 3x^3 - 18x^2 + 24x + 5x^2 - 30x + 40$$
$$= 3x^3 - 13x^2 - 6x + 40$$

49. $(x + 4)(2x - 7)(x - 4)$
$$= (x + 4)(x - 4)(2x - 7)$$
$$= (x^2)(2x - 7) - 16(2x - 7)$$
$$= 2x^3 - 7x^2 - 32x + 112$$

51. $(x^2 + 2x + 1)(x^2 - x - 3)$
$$= x^2(x^2 - x - 3) + 2x(x^2 - x - 3) + 1(x^2 - x - 3)$$
$$= x^4 - x^3 - 3x^2 + 2x^3 - 2x^2 - 6x + x^2 - x - 3$$
$$= x^4 + x^3 - 4x^2 - 7x - 3$$

53. $(x^2 + 3y)(x^2 - 3y)(2x^2 + 5xy - 6y^2)$
$$= (x^4 - 9y^2)(2x^2 + 5xy - 6y^2)$$
$$= x^4(2x^2 + 5xy - 6y^2) - 9y^2(2x^2 + 5xy - 6y^2)$$
$$= 2x^6 + 5x^5 y - 6x^4 y^2 - 18x^2 y^2 - 45xy^3 + 54y^4$$

Cumulative Review Problems

55. Let x = amount invested at 7%

$18,000 - x$ = amount invested at 11%

$0.07x + 0.11(18,000 - x) = 1540$

$7x + 11(18,000 - x) = 154,000$

$7x + 198,000 - 11x = 154,000$

$-4x = -44,000$

$x = 11,000$

$18,000 - x = 7000$

She invested $11,000 at 7% and $7000 at 11%.

Putting Your Skills to Work

1. Area Shaded = area of large square (-)

   Area of small square

   $A = (x+y)^2 - y^2 = x^2 + 2xy + y^2 - y^2$

   $\qquad = x^2 + 2xy$

3. $x = 25 - 12 = 13$

   $y = 12$

   $A = x^2 + 2xy = 13^2 + 2(13)(12) = 481 \text{ in}^2$

5. $A = x^2 - y^2$

   $x = 8, \quad y = 3, \quad A = 8^2 - 3^2 = 55 \text{ in}^2$

   $x = 25, \quad y = 12, \quad A = 25^2 - 12^2 = 481 \text{ in}^2$

7. $x = 15, \quad y = 6$

   $A = (x+y)(x-y) = (15+6)(15-6) = 189 \text{ in}^2$

Exercises 4.6

1. $\dfrac{8x^3 - 20x^2 + 4x}{2x} = \dfrac{8x^3}{2x} - \dfrac{20x^2}{2x} + \dfrac{4x}{2x}$

   $\qquad = 4x^2 - 10x + 2$

3. $\dfrac{8y^4 - 12y^3 - 4y^2}{4y^2} = \dfrac{8y^4}{4y^2} - \dfrac{12y^3}{4y^2} - \dfrac{4y^2}{4y^2}$

   $\qquad = 2y^2 - 3y - 1$

5. $\dfrac{49x^6 - 21x^4 + 56x^2}{7x^2} = \dfrac{49x^6}{7x^2} - \dfrac{21x^4}{7x^2} + \dfrac{56x^2}{7x^2}$

   $\qquad = 7x^4 - 3x^2 + 8$

7. $(16x^5 - 72x^3 + 24x^2) \div 8x^2$

   $= \dfrac{16x^5}{8x^2} - \dfrac{72x^3}{8x^2} + \dfrac{24x^2}{8x^2}$

   $= 2x^3 - 9x + 3$

9. $(30x^4 - 21x^3 - 3x^2 + 15x) \div (3x)$

   $= \dfrac{30x^4}{3x} - \dfrac{21x^3}{3x} - \dfrac{3x^2}{3x} + \dfrac{15x}{3x}$

   $= 10x^3 - 7x^2 - x + 5$

11.
$$
\begin{array}{r}
3x + 5 \\
2x+1{\overline{\smash{\big)}\,6x^2 + 13x + 5}} \\
\underline{6x^2 + 3x} \\
10x + 5 \\
\underline{10x + 5} \\
0
\end{array}
$$

$\dfrac{6x^2 + 13x + 5}{2x + 1} = 3x + 5$

Check: $(3x+5)(2x+1)$

$= 6x^2 + 3x + 10x + 5$

$= 6x^2 + 13x + 5$

13.
$$
\begin{array}{r}
x - 2 \\
x-7{\overline{\smash{\big)}\,x^2 - 9x - 6}} \\
\underline{x^2 - 7x} \\
-2x - 6 \\
\underline{-2x + 14} \\
-20
\end{array}
$$

$\dfrac{x^2 - 9x - 6}{x - 7} = x - 2 - \dfrac{20}{x - 7}$

Check: $(x-7)\left(x - 2 - \dfrac{20}{x-7}\right)$

$= x^2 - 2x - 7x + 14 - 20$

$= x^2 - 9x - 6$

71

15. $x+1{\overline{\smash{\big)}\,3x^3-\ x^2+4x-2}}$ with quotient $3x^2-4x+8$

$$\begin{array}{r} 3x^2-4x+8 \\ x+1{\overline{\smash{\big)}\,3x^3-\ x^2+4x-2}} \\ \underline{3x^3+3x^2} \\ -4x^2+4x-2 \\ \underline{-4x^2-4x} \\ 8x-2 \\ \underline{8x+8} \\ -10 \end{array}$$

$$\frac{3x^3-x^2+4x-2}{x+1}=3x^2-4x+8-\frac{10}{x+1}$$

Check: $(x+1)\left(3x^2-4x+8-\dfrac{10}{x+1}\right)$

$=3x^3-x^2+4x+8-10$

$=3x^3-x^2+4x-2$

17.
$$\begin{array}{r} x^2-x+2 \\ 2x+1{\overline{\smash{\big)}\,2x^3-x^2+3x+2}} \\ \underline{2x^3+x^2} \\ -2x^2+3x+2 \\ \underline{-2x^2-\ x} \\ 4x+2 \\ \underline{4x+2} \\ 0 \end{array}$$

$$\frac{2x^3-x^2+3x+2}{2x+1}=x^2-x+2$$

Check: $(2x+1)(x^2-x+2)$

$=2x^3-2x^2+4x+x^2-x+2$

$=2x^3-x^2+3x+2$

19.
$$\begin{array}{r} 2x^2+x-2 \\ 3x-1{\overline{\smash{\big)}\,6x^3+\ x^2-7x+2}} \\ \underline{6x^3-2x^2} \\ 3x^2-7x+2 \\ \underline{3x^2-\ x} \\ -6x+2 \\ \underline{-6x+2} \\ 0 \end{array}$$

$$\frac{6x^3+x^2-7x+2}{3x-1}=2x^2+x-2$$

21.
$$\begin{array}{r} 2y^2-9y+10 \\ 4y-1{\overline{\smash{\big)}\,8y^3-38y^2+49y+15}} \\ \underline{8y^3-\ 2y^2} \\ -36y^2+49y+15 \\ \underline{-36y^2+\ 9y} \\ 40y+15 \\ \underline{40y-10} \\ 25 \end{array}$$

$$\frac{8y^3-38y^2+49y+15}{4y-1}=2y^2-9y+10+\frac{25}{4y-1}$$

23.
$$\begin{array}{r} y^2-4y-1 \\ y+3{\overline{\smash{\big)}\,y^3-\ y^2-13y-12}} \\ \underline{y^3+3y^2} \\ -4y^2-13y-12 \\ \underline{-4y^2-12y} \\ -y-12 \\ \underline{-y-\ 3} \\ -9 \end{array}$$

$$(y^3-y^2-13y-12)\div(y+3)=y^2-4y-1-\frac{9}{y+3}$$

25.
$$\begin{array}{r} 2y^2+y+2 \\ 2y-1{\overline{\smash{\big)}\,4y^3+0y^2+3y-1}} \\ \underline{4y^3-2y^2} \\ 2y^2+3y-1 \\ \underline{2y^2-\ y} \\ 4y-1 \\ \underline{4y-2} \\ 1 \end{array}$$

$$(4y^3-1+3y)\div(2y-1)=2y^2+y+2+\frac{1}{2y-1}$$

27.
$$\begin{array}{r} y^3-3y^2+6y-16 \\ y+3{\overline{\smash{\big)}\,y^4+0y^3-3y^2+2y-1}} \\ \underline{y^4+3y^3} \\ -3y^3-3y^2+2y-1 \\ \underline{-3y^3-9y^2} \\ 6y^2+2y-1 \\ \underline{6y^2+18y} \\ -16y-1 \\ \underline{-16y-48} \\ 47 \end{array}$$

$$(y^4-3y^2+2y-1)\div(y+3)$$

$$=y^3-3y^2+6y-16+\frac{47}{y+3}$$

72

$$y - 2 \overline{\smash{\big)}\, y^4 + 0y^3 - 9y^2 + 0y - 5}$$

29. Long division:

$$
\begin{array}{r}
y^3 + 2y^2 - 5y - 10 \\
y - 2 \overline{\smash{\big)}\, y^4 + 0y^3 - 9y^2 + 0y - 5} \\
\underline{y^4 - 2y^3} \\
2y^3 - 9y^2 + 0y - 5 \\
\underline{2y^3 - 4y^2} \\
-5y^2 + 0y - 5 \\
\underline{-5y^2 + 10y} \\
-10y - 5 \\
\underline{-10y + 20} \\
-25
\end{array}
$$

$$\left(y^4 - 9y^2 - 5\right) \div \left(y - 2\right) = y^3 + 2y^2 - 5y - 10 - \frac{25}{y-2}$$

Cumulative Review Problems

31. $x - 6 \le 4(x - 3) - x$

$\quad x - 6 \le 4x - 12 - x$

$\quad -2x \le -6$

$\quad x \ge 3$

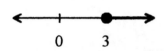

$\qquad$ 0 $\qquad$ 3

33. $3 + 2x > 6x - 9 - 7x + 5x$

$\quad 3 + 2x > 4x - 9$

$\quad -2x > -12$

$\quad x < 6$

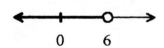

$\qquad$ 0 $\qquad$ 6

Chapter 4 - Review Problems

1. $\left(-6a^2\right)\left(3a^5\right) = -18a^7$

3. $\left(3xy^2\right)\left(2x^3y^4\right) = 6x^4y^6$

5. $\dfrac{8^{20}}{8^3} = 8^{17}$

7. $\dfrac{x^8}{x^{12}} = \dfrac{1}{x^4}$

9. $\dfrac{3x^8 y^0}{9x^4} = \dfrac{x^4}{3}$

11. $\dfrac{-12a^3b^6}{18a^2b^{12}} = -\dfrac{2a}{3b^6}$

13. $\left(x^3\right)^8 = x^{3(8)} = x^{24}$

15. $\left(-3a^3b^2\right)^2 = (-3)^2 a^{3(2)} b^{2(2)} = 9a^6b^4$

17. $\dfrac{2x^4}{3y^2} = \dfrac{2x^4}{3y^2}$

19. $\dfrac{\left(2xy^2\right)^3}{\left(2xy^2\right)^4} = \dfrac{1}{\left(2xy^2\right)^1} = \dfrac{1}{2xy^2}$

21. $x^{-3} = \dfrac{1}{x^3}$

23. $\dfrac{2x^{-6}}{y^{-3}} = \dfrac{2y^3}{x^6}$

25. $\left(2x^3\right)^{-2} = \dfrac{1}{\left(2x^3\right)^2} = \dfrac{1}{2^2 x^{3(2)}} = \dfrac{1}{4x^6}$

27. $\dfrac{4x^{-5}y^{-6}}{w^{-2}z^8} = \dfrac{4w^2}{x^5y^6z^8}$

29. $156{,}340{,}200{,}000 = 1.563402 \times 10^{11}$

31. $179{,}632 = 1.79632 \times 10^5$

33. $0.00006173 = 6.173 \times 10^{-5}$

35. $1.2 \times 10^5 = 120{,}000$

37. $3 \times 10^6 = 3{,}000{,}000$

39. $5.708 \times 10^{-8} = 0.00000005708$

41. $\dfrac{(28{,}000{,}000)(5{,}000{,}000{,}000)}{7{,}000}$

$\quad = \dfrac{\left(2.8 \times 10^7\right)\left(5 \times 10^9\right)}{7 \times 10^3}$

$\quad = \dfrac{14 \times 10^{16}}{7 \times 10^3} = 2 \times 10^{13}$

43. $\left(1.6 \times 10^{-3}\right)\left(3.0 \times 10^{-5}\right)\left(2.0 \times 10^{-2}\right)$

$\quad = 9.6 \times 10^{-10}$

45. $r = 40,000 = 4 \times 10^4$

    $t = 365(24) = 8760 = 8.76 \times 10^3$

    $d = rt = \left(4 \times 10^4\right)\left(8.76 \times 10^3\right)$

        $= 35.04 \times 10^7$

        $= 3.504 \times 10^8 \text{ km}$

47. $\dfrac{60}{1 \times 10^{-8}} = 60 \times 10^8 = 6 \times 10^9$

    In one minute the computer can
    perform 6×10^9 operations.

49. $\left(-7x^2 - 6x + 3\right) + \left(2x^2 - 10x - 15\right)$

    $= (-7 + 2)x^2 + (-6 - 10)x + 3 - 15$

    $= -5x^2 - 16x - 12$

51. $\left(4x^3 - x^2 - x + 3\right) - \left(-3x^3 + 2x^2 + 5x - 1\right)$

    $= \left(4x^3 - x^2 - x + 3\right) + \left(3x^3 - 2x^2 - 5x + 1\right)$

    $= (4 + 3)x^3 + (-1 - 2)x^2 + (-1 - 5)x + 3 + 1$

    $= 7x^3 - 3x^2 - 6x + 4$

53. $\left(8x^2 + 2y + 3xy\right) + \left(-5x^2 - 8y + 17xy\right)$

    $= (8 - 5)x^2 + (2 - 8)y + (3 + 17)xy$

    $= 3x^2 - 6y + 20xy$

55. $\left(4x^2y^2 + 6xy - 2y\right) - \left(-7x^2y^2 + 5xy + 3y^2\right)$

    $= \left(4x^2y^2 + 6xy - 2y\right) + \left(7x^2y^2 - 5xy + 3y^2\right)$

    $= (4 + 7)x^2y^2 + (6 - 5)xy - 2y - 3y^2$

    $= 11x^2y^2 + xy - 2y - 3y^2$

57. $\left(2x^2 - 7\right) - \left(3x^2 - 4\right) + \left(-5x^2 - 6x\right)$

    $= \left(2x^2 - 7\right) + \left(-3x^2 + 4\right) + \left(-5x^2 - 6x\right)$

    $= (2 - 3 - 5)x^2 - 6x - 7 + 4$

    $= -6x^2 - 6x - 3$

59. $(8x - 2)(4x - 3) = 32x^2 - 24x - 8x + 6$

    $= 32x^2 - 32x + 6$

61. $5x\left(2x^2 - 6x + 3\right) = 10x^3 - 30x^2 + 15x$

63. $\left(x^3 - 3x^2 + 5x - 2\right)(4x) = 4x^4 - 12x^3 + 20x^2 - 8x$

65. $\left(2x^2 - 3\right)\left(4x^2 - 5y\right) = 8x^4 - 10x^2y - 12x^2 + 15y$

67. $(3x - 2)^2 = 9x^2 + 2(-2)(3)x + 4$

    $= 9x^2 - 12x + 4$

69. $(7x + 6y)(7x - 6y) = (7x)^2 - (6y)^2 = 49x^2 - 36y^2$

71. $(8x + 9y)^2 = 64x^2 + 2(8x)(9y) + 81y^2$

    $= 64x^2 + 144xy + 81y^2$

73. $\left(x^3 + 2x^2 - x - 4\right)(2x - 3)$

    $= 2x\left(x^3 + 2x^2 - x - 4\right) - 3\left(x^3 + 2x^2 - x - 4\right)$

    $= 2x^4 + 4x^3 - 2x^2 - 8x - 3x^3 - 6x^2 + 3x + 12$

    $= 2x^4 + x^3 - 8x^2 - 5x + 12$

75. $(2x + 3)^3$

    $= (2x + 3)(2x + 3)^2$

    $= (2x + 3)\left(4x^2 + 12x + 9\right)$

    $= 2x\left(4x^2 + 12x + 9\right) + 3\left(4x^2 + 12x + 9\right)$

    $= 8x^3 + 24x^2 + 18x + 12x^2 + 36x + 27$

    $= 8x^3 + 36x^2 + 54x + 27$

77. $\left(30x^5 + 35x^4 - 90x^3\right) \div \left(5x^2\right)$

    $= \dfrac{30x^5}{5x^2} + \dfrac{35x^4}{5x^2} - \dfrac{90x^3}{5x^2}$

    $= 6x^3 + 7x^2 - 18x$

79. $\left(106x^6 - 24x^5 + 38x^4 + 26x^3\right) \div \left(2x^3\right)$

    $= \dfrac{106x^6}{2x^3} - \dfrac{24x^5}{2x^3} + \dfrac{38x^4}{2x^3} + \dfrac{26x^3}{2x^3}$

    $= 53x^3 - 12x^2 + 19x + 13$

81. $\begin{array}{r} 4x + 1 \\ 4x - 3 \overline{)\, 16x^2 - 8x - 3} \\ \underline{-16x^2 + 12x} \\ 4x - 3 \\ \underline{4x - 3} \end{array}$

    $\left(16x^2 - 8x - 3\right) \div (4x - 3) = 4x + 1$

74

83.

$$
\begin{array}{r}
3x^2 + 2x + 4 \\
2x-1{\overline{\smash{\big)}\,6x^3 + x^2 + 6x + 5}} \\
\underline{6x^3 - 3x^2} \\
4x^2 + 6x \\
\underline{4x^2 - 2x} \\
8x + 5 \\
\underline{8x - 4} \\
9
\end{array}
$$

$$\left(6x^3 + x^2 + 6x + 5\right) \div \left(2x - 1\right)$$

$$= 3x^2 + 2x + 4 + \frac{9}{2x-1}$$

85.

$$
\begin{array}{r}
4x + 1 \\
3x+2{\overline{\smash{\big)}\,12x^2 + 11x + 2}} \\
\underline{12x^2 + 8x} \\
3x + 2 \\
\underline{3x + 2}
\end{array}
$$

$$\left(12x^2 + 11x + 2\right) \div \left(3x + 2\right) = 4x + 1$$

87.

$$
\begin{array}{r}
2x^2 + 4x + 5 \\
x-2{\overline{\smash{\big)}\,2x^3 + 0x^2 - 3x + 1}} \\
\underline{2x^3 - 4x^2} \\
4x^2 - 3x \\
\underline{4x^2 - 8x} \\
5x + 1 \\
\underline{5x - 10} \\
11
\end{array}
$$

$$\left(2x^3 - 3x + 1\right) \div \left(x - 2\right) = 2x^2 + 4x + 5 + \frac{11}{x-2}$$

89.

$$
\begin{array}{r}
4x^2 + 5x + 1 \\
3x-1{\overline{\smash{\big)}\,12x^3 + 11x^2 - 2x - 1}} \\
\underline{12x^3 - 4x^2} \\
15x^2 - 2x \\
\underline{15x^2 - 5x} \\
3x - 1 \\
\underline{3x - 1}
\end{array}
$$

$$\left(12x^3 + 11x^2 - 2x - 1\right) \div \left(3x - 1\right) = 4x^2 + 5x + 1$$

Test - Chapter 4

1. $\left(3^{10}\right)\left(3^{24}\right) = 3^{10+24} = 3^{34}$

3. $\left(8^4\right)^6 = 8^{4(6)} = 8^{24}$

5. $\dfrac{-35x^8 y^{10}}{25x^5 y^{10}} = -\dfrac{7x^3}{5}$

7. $\left(\dfrac{7a^7 b^2}{3c^0}\right)^2 = \dfrac{7^2 a^{7(2)} b^{2(2)}}{3^2 c^{0(2)}} = \dfrac{49a^{14} b^4}{9}$

9. $5^{-3} = \dfrac{1}{5^3} = \dfrac{1}{125}$

11. $\dfrac{2x^{-3} y^{-4}}{w^{-6} z^8} = \dfrac{2w^6}{x^3 y^4 z^8}$

13. $5.82 \times 10^8 = 582{,}000{,}000$

15. $\left(2x^2 - 3x - 6\right) + \left(-4x^2 + 8x + 6\right)$
$= (2-4)x^2 + (-3+8)x + 6 - 6$
$= -2x^2 + 5x$

17. $-7x^2\left(3x^3 - 4x^2 + 6x - 2\right)$
$= -21x^5 + 28x^4 - 42x^3 + 14x^2$

19. $(5a - 4b)(2a + 3b)$
$= 10a^2 + 15ab - 8ab - 12b^2$
$= 10a^2 + 7ab - 12b^2$

21. $\left(7x^2 + 2y^2\right)^2$
$= 49x^4 + 2\left(7x^2\right)\left(2y^2\right) + 4y^4$
$= 49x^4 + 28x^2 y^2 + 4y^4$

23. $(3x - 2)\left(4x^3 - 2x^2 + 7x - 5\right)$
$= 3x\left(4x^3 - 2x^2 + 7x - 5\right) - 2\left(4x^3 - 2x^2 + 7x - 5\right)$
$= 12x^4 - 6x^3 + 21x^2 - 15x - 8x^3 + 4x^2 - 14x + 10$
$= 12x^4 - 14x^3 + 25x^2 - 29x + 10$

25. $\dfrac{15x^6 - 5x^4 + 25x^3}{5x^3} = \dfrac{15x^6}{5x^3} - \dfrac{5x^4}{5x^3} + \dfrac{25x^3}{5x^3}$
$= 3x^3 - x + 5$

Cumulative Test - Chapters 0 - 4

1. $\dfrac{5}{12} - \dfrac{7}{8} = \dfrac{10}{24} - \dfrac{21}{24} = -\dfrac{11}{24}$

3. $\left(-4\frac{1}{2}\right) \div \left(5\frac{1}{4}\right)$

$= \left(-\frac{9}{2}\right) \div \left(\frac{21}{4}\right)$

$= \left(-\frac{3 \cdot 3}{2}\right)\left(\frac{2 \cdot 2}{3 \cdot 7}\right)$

$= -\frac{6}{7}$

5. $7x(3x-4) - 5x(2x-3) - (3x)^2$

$= 21x^2 - 28x - 10x^2 + 15x - 9x^2$

$= 2x^2 - 13x$

7. $7x - 3(4-2x) = 14x - (3-x)$

$7x - 12 + 6x = 14x - 3 + x$

$13x - 12 = 15x - 3$

$-9 = 2x$

$-\frac{9}{2} = x$

9. $4 - 7x < 11$

$-7x < 7$

$x > -1$

11. $\dfrac{11{,}904}{x} = 0.96$

$11{,}904 = 0.96x$

$12{,}400 = x$

VBM has 12,400 employees

13. $(3x-7)(5x-4)$

$= 15x^2 - 12x - 35x + 28$

$= 15x^2 - 47x + 28$

15. $(2x+1)^3$

$= (2x+1)(2x+1)^2$

$= (2x+1)(4x^2 + 4x + 1)$

$= 2x(4x^2 + 4x + 1) + 1(4x^2 + 4x + 1)$

$= 8x^3 + 8x^2 + 2x + 4x^2 + 4x + 1$

$= 8x^3 + 12x^2 + 6x + 1$

17. $\dfrac{14x^8 y^3}{-21x^5 y^{12}} = -\dfrac{2x^3}{3y^9}$

19. $\dfrac{9x^{-3}y^{-4}}{w^2 z^{-8}} = \dfrac{9z^8}{x^3 y^4 w^2}$

21. $\dfrac{(2.0 \times 10^{-12})(8.0 \times 10^{-20})}{4.0 \times 10^3} = \dfrac{16.0 \times 10^{-32}}{4.0 \times 10^3}$

$= 4.0 \times 10^{-35}$

23. $-6xy^2(6x^2 - 3xy + 8y^2)$

$= -36x^3 y^2 + 18x^2 y^3 - 48xy^4$

Practice Quiz: Sections 4.1 - 4.3

Find the following and leave your answer in exponential form.

1. $(4^7)(4^{18})$

2. $\dfrac{4a^3}{2a^{10}}$

3. $(y^6)^8$

4. Use positive exponents to rewrite

$\dfrac{-3a^{-2}}{b^3 c^{-5}}$

5. Combine

$(3x^2 - 5x + 8) - (-2x^2 - 3x - 4)$

Practice Quiz: Sections 4.4 - 4.6

Multiply the following polynomials.

1. $-5x(3x^2 - 7x)$

2. $(2x - 3y)(x + 4y)$

3. $(2x + 5)(2x - 5)$

4. $(5a - 7)^2$

Divide the following polynomials.

5. $(12x^3 - 6x^2 + 8x) \div 2x$

6. $(2x^2 - 5x + 2) \div (2x - 1)$

76

<u>Answers to Practice Quiz</u>

Sections 4.1 - 4.3

1. 4^{25}

2. $\dfrac{2}{a^7}$

3. y^{48}

4. $\dfrac{-3c^5}{a^2 b^3}$

5. $5x^2 - 2x + 12$

<u>Answers to Practice Quiz</u>

Sections 4.4 - 4.6

1. $-15x^3 + 35x^2$

2. $2x^2 + 5xy - 12y^2$

3. $4x^2 - 25$

4. $25a^2 - 70a + 49$

5. $6x^2 - 3x + 4$

6. $x - 2$

1. $2x^2 - 6xy + 12xy^2 = 2x(x - 3y + 6y^2)$

3. $36ab^2 - 18ab = 18ab(2b - 1)$

5. $3x^2 - 4y + 3xy - 4x = 3x^2 - 4x + 3xy - 4y$
$$= x(3x - 4) + y(3x - 4)$$
$$= (3x - 4)(x + y)$$

7. $x^2 - 22x - 48 = (x - 24)(x + 2)$

9. $x^2 + 6x + 5 = (x + 5)(x + 1)$

11. $3x^2 - 6x - 189 = 3(x^2 - 2x - 63)$
$$= 3(x - 9)(x + 7)$$

13. $6y^2 + 5yz - 6z^2 = (2y + 3z)(3y - 2z)$

15. $81x^4 - 16 = (9x^2)^2 - (4)^2$
$$= (9x^2 + 4)(9x^2 - 4)$$
$$= (9x^2 + 4)(3x + 2)(3x - 2)$$

17. $25x^2 + 80x + 64 = (5x + 8)^2$

19. $32x^2 y^2 - 48xy^2 + 18y^2 = 2y^2(16x^2 - 24x + 9)$
$$= 2y^2(4x - 3)^2$$

21. $2x^2 + x - 3 = 0$
$(2x + 3)(x - 1) = 0$
$2x + 3 = 0 \qquad x - 1 = 0$
$$x = -\frac{3}{2} \qquad x = 1$$
Roots: $-\dfrac{3}{2}$, 1

23. $\dfrac{3x^2 - 7x}{2} = 3$
$3x^2 - 7x = 6$
$3x^2 - 7x - 6 = 0$
$(3x + 2)(x - 3) = 0$

$3x + 2 = 0 \qquad x - 3 = 0$
$$x = -\frac{2}{3} \qquad x = 3$$
Roots: $-\dfrac{2}{3}$, 3

Exercises 5.1

1. $3x^2$ and $5x^3$ are called <u>factors</u>.

3. The factoring is not complete because $6a^3 + 3a^2 - 9a$ contains a common factor of $3a$.

5. $5ab - 5b = 5b(a - 1)$

7. $18wz - 27w^2 z = 9wz(2 - 3w)$

9. $8x^3 - 10x^2 - 14x = 2x(4x^2 - 5x - 7)$

11. $9pqr - 12p^2 qr - 14rx$
$$= r(9pq - 12p^2 q - 14x)$$

13. $a^4 y - a^3 y^2 + a^2 y^3 - 2a$
$$= a(a^3 y - a^2 y^2 + ay^3 - 2)$$

15. $6x^9 - 8x^7 + 4x^5 = 2x^5(3x^4 - 4x^2 + 2)$

17. $9a^2 b^2 - 36ab = 9ab(ab - 4)$

19. $40a^2 - 16ab - 24a = 8a(5a - 2b - 3)$

21. $34x^2 - 51xy - 17x = 17x(2x - 3y - 1)$

23. $18xy^2 + 6x^2 y + 24xy - 12x^2 y^2$
$$= 6xy(3y + x + 4 - 2xy)$$

25. $5(x - 2y) - z(x - 2y) = (x - 2y)(5 - z)$

27. $5x(x - 7) + 3(x - 7) = (x - 7)(5x + 3)$

29. $5b(c + 2d) - 2a(c + 2d) = (c + 2d)(5b - 2a)$

31. $5a(bc - 1) + b(bc - 1) + c(bc - 1)$
$$= (bc - 1)(5a + b + c)$$

33. $4(x^2 + 8) - 2y(x^2 + 8) - w(x^2 + 8)$
$$= (x^2 + 8)(4 - 2y - w)$$

35. $3x^2(x-2y)-(x-2y)=(x-2y)(3x^2-1)$

37. $2ab(9x+5y)+(9x+5y)=(9x+5y)(2ab+1)$

39. $3x^2y(5a-3b+1)-19xy(5a-3b+1)-y(5a-3b+1)$
$= (5a-3b+1)(3x^2y-19xy-y)$
$= y(5a-3b+1)(3x^2-19x-1)$

41. $173{,}946x^2 - 347{,}892y^2 - 637{,}802z^2$
$= 57{,}982(3x^2-6y^2-11z^2)$

43. $C =$ Total cost
$C = 25.95a + 25.95b + 25.95c + 25.95d$
$= \$25.95(a+b+c+d)$

Cumulative Review Problems

45. $(4a-b)^2 = (4a)^2 + 2(4a)(-b)+(-b)^2$
$= 16a^2 - 8ab + b^2$

47. $\quad x =$ last year's cost
$0.08x =$ increase in cost
$x + 0.08x = 13{,}284$
$1.08x = 13{,}284$
$x = 12{,}300$

Last year's cost was \$12,300

Exercises 5.2

1. $3x-6+xy-2y = 3(x-2)+y(x-2)$
$= (x-2)(3+y)$

Check: $(x-2)(3+y) = 3x + xy - 6 - 2y$
$= 3x - 6 + xy - 2y$

3. $ax+ay-2bx-2by = a(x+y)-2b(x+y)$
$= (x+y)(a-2b)$

Check: $(x+y)(a-2b) = ax - 2bx + ay - 2by$
$= ax + ay - 2bx - 2by$

5. $x^3 - 4x^2 + 3x - 12 = x^2(x-4)+3(x-4)$
$= (x-4)(x^2+3)$

Check: $(x-4)(x^2+3) = x^3 + 3x - 4x^2 - 12$
$= x^3 - 4x^2 + 3x - 12$

7. $3ax + bx - 6a - 2b = x(3a+b)-2(3a+b)$
$= (3a+b)(x-2)$

Check: $(3a+b)(x-2) = 3ax - 6a + bx - 2b$
$= 3ax + bx - 6a - 2b$

9. $5a + 12bc + 10b + 6ac = 5a + 10b + 6ac + 12bc$
$= 5(a+2b)+6c(a+2b)$
$= (a+2b)(5+6c)$

Check: $(a+2b)(5+6c) = 5a + 6ac + 10b + 12bc$
$= 5a + 12bc + 10b + 6ac$

11. $5a - 5b - 2ax + 2xb = 5(a-b)-2x(a-b)$
$= (a-b)(5-2x)$

Check: $(a-b)(5-2x) = 5a - 2ax - 5b + 2bx$
$= 5a - 5b - 2ax + 2xb$

13. $y^2 - 2y - 3y + 6 = y(y-2)-3(y-2)$
$= (y-2)(y-3)$

15. $3ax - y + 3ay - x = 3ax + 3ay - x - y$
$= 3a(x+y)-1(x+y)$
$= (x+y)(3a-1)$

17. $6ax - y + 2ay - 3x = 6ax - 3x + 2ay - y$
$= 3x(2a-1)+y(2a-1)$
$= (2a-1)(3x+y)$

19. $x^2 + 6x - 8x - 48 = x(x+6)-8(x+6)$
$= (x+6)(x-8)$

21. $28x^2 + 8xy^2 + 21xw + 6y^2w$
$= 4x(7x+2y^2)+3w(7x+2y^2)$
$= (7x+2y^2)(4x+3w)$

23. $8a^2 - 6ab + 20ae - 15be$
$= 2a(4a-3b)+5e(4a-3b)$
$= (4a-3b)(2a+5e)$

25. $44x^3 - 63y^2w - 36xy^2 + 77x^2w$

$= 44x^3 + 77x^2w - 36xy^2 - 63y^2w$

$= 11x^2(4x + 7w) - 9y^2(4x + 7w)$

$= (4x + 7w)(11x^2 - 9y^2)$

27. Rearrange the terms so that factoring gives the same expression in parentheses.

$6a^2 - 12bd - 8ad + 9ab$

$= 6a^2 - 8ad + 9ab - 12bd$

$= 2a(3a - 4d) + 3b(3a - 4d)$

$= (3a - 4d)(2a + 3b)$

Cumulative Review Problems

29. A polynomial of three terms is called a <u>trinomial</u>.

31. $x =$ score on final

$$\frac{79 + 95 + 82 + 89 + 2x}{6} = 88$$

$$345 + 2x = 528$$

$$2x = 183$$

$$x = 91.5$$

He must score 91.5

Exercises 5.3

1. Find two numbers whose <u>product</u> is 6 and whose <u>sum</u> is 5.

3. $x^2 + 9x + 8$; Product: 8, Sum: 9, + signs

$x^2 + 9x + 8 = (x + 1)(x + 8)$

5. $x^2 + 8x + 12$; Product: 12, Sum: 8, + signs

$x^2 + 8x + 12 = (x + 2)(x + 6)$

7. $x^2 - 4x + 3$; Product: 3, Sum: -4, - signs

$x^2 - 4x + 3 = (x - 1)(x - 3)$

9. $x^2 - 13x + 22$; Product: 22, Sum: -13, - signs

$x^2 - 13x + 22 = (x - 2)(x - 11)$

11. $x^2 + x - 12$; Product: -12, Sum: 1,
Opposite signs with larger absolute value +.

$x^2 + x - 12 = (x + 4)(x - 3)$

13. $x^2 - 13x - 14$; Product: -14, Sum: -13,
Opposite signs with larger absolute value - .

$x^2 - 13x - 14 = (x - 14)(x + 1)$

15. $x^2 + x - 20$; Product: -20, Sum: 1
Opposite signs with larger absolute value +.

$x^2 + x - 20 = (x + 5)(x - 4)$

17. $x^2 - 5x - 24$; Product: -24, Sum: -5
Opposite signs with larger absolute value -.

$x^2 - 5x - 24 = (x - 8)(x + 3)$

19. $x^2 - 7x + 10 = (x - 2)(x - 5)$

Check: $(x - 2)(x - 5) = x^2 - 5x - 2x + 10$

$= x^2 - 7x + 10$

21. $x^2 - x - 2 = (x - 2)(x + 1)$

Check: $(x - 2)(x + 1) = x^2 + x - 2x - 2$

$= x^2 - x - 2$

23. $x^2 + 5x - 14 = (x + 7)(x - 2)$

Check: $(x + 7)(x - 2) = x^2 - 2x + 7x - 14$

$= x^2 + 5x - 14$

25. $x^2 - 9x + 20 = (x - 5)(x - 4)$

Check: $(x - 5)(x - 4) = x^2 - 4x - 5x + 20$

$= x^2 - 9x + 20$

27. $x^2 + 17x + 30 = (x + 2)(x + 15)$

Check: $(x + 2)(x + 15) = x^2 + 15x + 2x + 30$

$= x^2 + 17x + 30$

29. $y^2 - 4y - 5 = (y - 5)(y + 1)$

Check: $(y - 5)(y + 1) = y^2 + y - 5y - 5$

$= y^2 - 4y - 5$

31. $a^2 + 6a - 16 = (a + 8)(a - 2)$

Check: $(a + 8)(a - 2) = a^2 - 2a + 8a - 16$

$= a^2 + 6a - 16$

33. $x^2 - 12x + 32 = (x-4)(x-8)$

Check: $(x-4)(x-8) = x^2 - 8x - 4x + 32$
$= x^2 - 12x + 32$

35. $x^2 + 4x - 21 = (x+7)(x-3)$

Check: $(x+7)(x-3) = x^2 - 3x + 7x - 21$
$= x^2 + 4x - 21$

37. $x^2 + 13x + 40 = (x+5)(x+8)$

Check: $(x+5)(x+8) = x^2 + 8x + 5x + 40$
$= x^2 + 13x + 40$

39. $x^2 - 21x - 22 = (x-22)(x+1)$

Check: $(x-22)(x+1) = x^2 + x - 22x - 22$
$= x^2 - 21x - 22$

41. $x^2 + 9x - 36 = (x+12)(x-3)$

Check: $(x+12)(x-3) = x^2 - 3x + 12x - 36$
$= x^2 + 9x - 36$

43. $x^2 - xy - 42y^2 = (x-7y)(x+6y)$

Check: $(x-7y)(x+6y) = x^2 + 6xy - 7xy - 42y^2$
$= x^2 - xy - 42y^2$

45. $x^2 - 16xy + 63y^2 = (x-9y)(x-7y)$

Check: $(x-9y)(x-7y) = x^2 - 7xy - 9xy + 63y^2$
$= x^2 - 16xy + 63y^2$

47. $x^4 - 3x^2 - 40 = (x^2-8)(x^2+5)$

Check: $(x^2-8)(x^2+5) = x^4 + 5x^2 - 8x^2 - 40$
$= x^4 - 3x^2 - 40$

49. $x^4 + 32x^2 + 60 = (x^2+2)(x^2+30)$

Check: $(x^2+2)(x^2+30) = x^4 + 30x^2 + 2x^2 + 60$
$= x^4 + 32x^2 + 60$

51. $2x^2 - 12x + 16 = 2(x^2 - 6x + 8)$
$= 2(x-2)(x-4)$

53. $3x^2 - 6x - 72 = 3(x^2 - 2x - 24)$
$3(x-6)(x+4)$

55. $4x^2 + 24x + 20 = 4(x^2 + 6x + 5)$
$4(x+1)(x+5)$

57. $5x^2 + 10x - 15 = 5(x^2 + 2x - 3)$
$= 5(x+3)(x-1)$

59. $6x^2 + 18x + 12 = 6(x^2 + 3x + 2)$
$= 6(x+2)(x+1)$

61. $2x^2 - 32x + 126 = 2(x^2 - 16x + 63)$
$= 2(x-9)(x-7)$

Cumulative Review Problems

63. $A = P + rPt$

$A - P = rPt$

$\dfrac{A-P}{rP} = t$

65. $c = $ car's speed

$c = \dfrac{d}{t} = \dfrac{d}{2}$

$c + 20 = $ train's speed

$\dfrac{d}{2} + 20 = \dfrac{d}{1.5}$

$1.5d + 60 = 2d$

$60 = 0.5d$

$120 = d$

It is 120 mi from Xanadu to Yreka.

Exercises 5.4

1. $3x^2 + 7x + 2$; Grouping number: 6

$3x^2 + 7x + 2 = 3x^2 + 6x + x + 2$
$= 3x(x+2) + (x+2)$
$= (x+2)(3x+1)$

Check: $(x+2)(3x+1) = 3x^2 + x + 6x + 2$
$= 3x^2 + 7x + 2$

3. $2x^2 - 5x + 2$; Grouping number: 4

$2x^2 - 5x + 2 = 2x^2 - 4x - x + 2$

$\qquad = 2x(x-2) - 1(x-2)$

$\qquad = (x-2)(2x-1)$

Check: $(x-2)(2x-1) = 2x^2 - x - 4x + 2$

$\qquad\qquad = 2x^2 - 5x + 2$

5. $2x^2 + 3x - 5$; Grouping number: -10

$2x^2 + 3x - 5 = 2x^2 + 5x - 2x - 5$

$\qquad = x(2x+5) - (2x+5)$

$\qquad = (2x+5)(x-1)$

Check: $(2x+5)(x-1) = 2x^2 - 2x + 5x - 5$

$\qquad\qquad = 2x^2 + 3x - 5$

7. $2x^2 - 5x - 3$; Grouping number: -6

$2x^2 - 5x - 3 = 2x^2 - 6x + x - 3$

$\qquad = 2x(x-3) + (x-3)$

$\qquad = (x-3)(2x+1)$

Check: $(x-3)(2x+1) = 2x^2 + x - 6x - 3$

$\qquad\qquad = 2x^2 - 5x - 3$

9. $5x^2 + 3x - 2$; Grouping number: -10

$5x^2 + 3x - 2 = 5x^2 + 5x - 2x - 2$

$\qquad = 5x(x+1) - 2(x+1)$

$\qquad = (x+1)(5x-2)$

Check: $(x+1)(5x-2) = 5x^2 - 2x + 5x - 2$

$\qquad\qquad = 5x^2 + 3x - 2$

11. $6x^2 - 13x + 6$; Grouping number: 36

$6x^2 - 13x + 6 = 6x^2 - 9x - 4x + 6$

$\qquad = 3x(2x-3) - 2(2x-3)$

$\qquad = (2x-3)(3x-2)$

Check: $(2x-3)(3x-2) = 6x^2 - 4x - 9x + 6$

$\qquad\qquad = 6x^2 - 13x + 6$

13. $2x^2 + 3x - 20$; Grouping number: -40

$2x^2 + 3x - 20 = 2x^2 + 8x - 5x - 20$

$\qquad = 2x(x+4) - 5(x+4)$

$\qquad = (x+4)(2x-5)$

Check: $(x+4)(2x-5) = 2x^2 - 5x + 8x - 20$

$\qquad\qquad = 2x^2 + 3x - 20$

15. $9x^2 + 9x + 2$; Grouping number: 18

$9x^2 + 9x + 2 = 9x^2 + 3x + 6x + 2$

$\qquad = 3x(3x+1) + 2(3x+1)$

$\qquad = (3x+1)(3x+2)$

Check: $(3x+1)(3x+2) = 9x^2 + 6x + 3x + 2$

$\qquad\qquad = 9x^2 + 9x + 2$

17. $6x^2 - 5x - 6$; Grouping number: -36

$6x^2 - 5x - 6 = 6x^2 - 9x + 4x - 6$

$\qquad = 3x(2x-3) + 2(2x-3)$

$\qquad = (2x-3)(3x+2)$

Check: $(2x-3)(3x+2) = 6x^2 + 4x - 9x - 6$

$\qquad\qquad = 6x^2 - 5x - 6$

19. $8x^2 - 6x + 1$; Grouping number: 8

$8x^2 - 6x + 1 = 8x^2 - 4x - 2x + 1$

$\qquad = 4x(2x-1) - (2x-1)$

$\qquad = (2x-1)(4x-1)$

Check: $(2x-1)(4x-1) = 8x^2 - 2x - 4x + 1$

$\qquad\qquad = 8x^2 - 6x + 1$

21. $4x^2 + 16x - 9 = 4x^2 + 18x - 2x - 9$

$\qquad = 2x(2x+9) - (2x+9)$

$\qquad = (2x+9)(2x-1)$

23. $9y^2 - 13y + 4 = 9y^2 - 9y - 4y + 4$

$\qquad = 9y(y-1) - 4(y-1)$

$\qquad = (y-1)(9y-4)$

25. $5a^2 - 13a - 6 = 5a^2 - 15a + 2a - 6$

$\qquad = 5a(a-3) + 2(a-3)$

$\qquad = (a-3)(5a+2)$

27. $12x^2 - 20x + 3 = 12x^2 - 18x - 2x + 3$

$\qquad = 6x(2x-3) - (2x-3)$

$\qquad = (2x-3)(6x-1)$

29. $15x^2 + 4x - 4 = 15x^2 + 10x - 6x - 4$
$$= 5x(3x + 2) - 2(3x + 2)$$
$$= (3x + 2)(5x - 2)$$

31. $10x^2 + 21x + 9 = 10x^2 + 15x + 6x + 9$
$$= 5x(2x + 3) + 3(2x + 3)$$
$$= (2x + 3)(5x + 3)$$

33. $16x^2 - 6x - 1 = 16x^2 - 8x + 2x - 1$
$$= 8x(2x - 1) + (2x - 1)$$
$$= (2x - 1)(8x + 1)$$

35. $2x^4 + 15x^2 - 8 = 2x^4 + 16x^2 - x^2 - 8$
$$= 2x^2(x^2 + 8) - (x^2 + 8)$$
$$= (x^2 + 8)(2x^2 - 1)$$

37. $4x^2 + 8xy - 5y^2 = 4x^2 + 10xy - 2xy - 5y^2$
$$= 2x(2x + 5y) - y(2x + 5y)$$
$$= (2x + 5y)(2x - y)$$

39. $5x^2 + 16xy - 16y^2 = 5x^2 + 20xy - 4xy - 16y^2$
$$= 5x(x + 4y) - 4y(x + 4y)$$
$$= (x + 4y)(5x - 4y)$$

41. $10x^2 + 22x + 12 = 2(5x^2 + 11x + 6)$
$$= 2(5x + 6)(x + 1)$$

43. $12x^2 - 24x + 9 = 3(4x^2 - 8x + 3)$
$$= 3(2x - 3)(2x - 1)$$

45. $18x^2 + 39x - 15 = 3(6x^2 + 13x - 5)$
$$= 3(3x - 1)(2x + 5)$$

47. $21x^2 + 60x - 9 = 3(7x^2 + 20x - 3)$
$$= 3(7x - 1)(x + 3)$$

49. $12x^2 + 16x - 35 = (2x + 5)(6x - 7)$

51. $20x^2 - 27x + 9 = (4x - 3)(5x - 3)$

Cumulative Review Problems

53. $7x - 3(4 - 2x) = 2(x - 3) - (5 - x)$
$$7x - 12 + 6x = 2x - 6 - 5 + x$$
$$13x - 12 = 3x - 11$$
$$10x = 1$$
$$x = \frac{1}{10}$$

55. $x = $ measure of 1st angle
$2x - 5 = $ measure of 2nd angle
$x + 3(2x - 5) = $ measure of 3rd angle

$$x + 2x - 5 + x + 3(2x - 5) = 180$$
$$x + 2x - 5 + x + 6x - 15 = 180$$
$$10x - 20 = 180$$
$$10x = 200$$
$$x = 20, \text{ also}$$
$$2x - 5 = 35, \text{ and}$$
$$x + 3(2x - 5) = 125$$

The angles measure 20^0, 35^0 and 125^0

Exercises 5.5

1. $9x^2 - 16 = (3x)^2 - (4)^2$
$$= (3x + 4)(3x - 4)$$

3. $1 - 100s^2 = (1)^2 - (10s)^2$
$$= (1 + 10s)(1 - 10s)$$

5. $25x^2 - 16 = (5x)^2 - (4)^2$
$$= (5x + 4)(5x - 4)$$

7. $4x^2 - 25 = (2x)^2 - (5)^2$
$$= (2x + 5)(2x - 5)$$

9. $36x^2 - 25 = (6x)^2 - (5)^2$
$$= (6x + 5)(6x - 5)$$

11. $1 - 49x^2 = (1)^2 - (7x)^2$
$$= (1 + 7x)(1 - 7x)$$

13. $81x^2 - 49y^2 = (9x)^2 - (7y)^2$
$$= (9x + 7y)(9x - 7y)$$

15. $25 - 121x^2 = (5)^2 - (11x)^2$
$$= (5 + 11x)(5 - 11x)$$

17. $81x^2 - 100y^2 = (9x)^2 - (10y)^2$
$$= (9x + 10y)(9x - 10y)$$

19. $25a^2 - 1 = (5a)^2 - (1)^2$
$$= (5a + 1)(5a - 1)$$

21. $49x^2 - 25y^2 = (7x)^2 - (5y)^2$
$$= (7x + 5y)(7x - 5y)$$

23. $49x^2 + 14x + 1 = (7x + 1)^2$

25. $y^2 - 6y + 9 = (y - 3)^2$

27. $9x^2 - 24x + 16 = (3x - 4)^2$

29. $49x^2 + 28x + 4 = (7x + 2)^2$

31. $x^2 + 10x + 25 = (x + 5)^2$

33. $9x^2 - 42x + 49 = (3x - 7)^2$

35. $25x^2 - 40x + 16 = (5x - 4)^2$

37. $81x^2 + 36xy + 4y^2 = (9x + 2y)^2$

39. $9x^2 - 42xy + 49y^2 = (3x - 7y)^2$

41. $16a^2 + 72ab + 81b^2 = (4a + 9b)^2$

43. $9x^4 - 6x^2y + y^2 = (3x^2 - y)^2$

45. $49x^2 + 70x + 9 = (7x + 1)(7x + 9)$

47. $x^6 - 25 = (x^3)^2 - (5)^2$
$$= (x^3 + 5)(x^3 - 5)$$

49. $x^{10} - 36y^{10} = (x^5)^2 - (6y^5)^2$
$$= (x^5 + 6y^5)(x^5 - 6y^5)$$

51. $4x^8 + 12x^4 + 9 = (2x^4 + 3)^2$

53. You cannot factor $9x^2 + 1$ because there are no combinations of the product of two binomials that give $9x^2 + 1$.

55. If $16y^2 - 56y + c$ is a perfect square trinomial
$$56 = 2(4)\sqrt{c}$$
$$7 = \sqrt{c}$$
$$c = 49$$
There is only one value.

57. $27x^2 - 3 = 3(9x^2 - 1)$
$$= 3\left[(3x)^2 - (1)^2\right]$$
$$= 3(3x + 1)(3x - 1)$$

59. $125x^2 - 5 = 5(25x^2 - 1)$
$$= 5\left[(5x)^2 - (1)^2\right]$$
$$= 5(5x + 1)(5x - 1)$$

61. $12x^2 - 36x + 27 = 3(4x^2 - 12x + 9)$
$$= 3(2x - 3)^2$$

63. $98x^2 + 84x + 18 = 2(49x^2 + 42x + 9)$
$$= 2(7x + 3)^2$$

65. $x^2 - 9x + 14 = (x - 7)(x - 2)$

67. $2x^2 + 5x - 3 = (2x - 1)(x + 3)$

69. $16x^2 - 121 = (4x + 11)(4x - 11)$

71. $25x^2 - 20x + 4 = (5x - 2)^2$

73. $3x^2 + 6x - 45 = 3(x^2 + 2x - 15)$
$$= 3(x + 5)(x - 3)$$

75. $7x^2 - 63 = 7(x^2 - 9)$
$$= 7(x + 3)(x - 3)$$

77. $5x^2 + 20x + 20 = 5(x^2 + 4x + 4)$
$$= 5(x + 2)^2$$

79. $2x^2 - 32x + 126 = 2(x^2 - 16x + 63)$
$$= 2(x - 9)(x - 7)$$

Cumulative Review Problems

81. $(x + 4)(2x + 3)(x - 2)$
$$= (2x^2 + 11x + 12)(x - 2)$$
$$= 2x^3 + 11x^2 + 12x - 4x^2 - 22x - 24$$
$$= 2x^3 + 7x^2 - 10x - 24$$

83.
$$\begin{array}{r} x^2 + 3x + 4 \\ x - 2\overline{)x^3 + x^2 - 2x - 11} \\ \underline{x^3 - 2x^2} \\ 3x^2 - 2x \\ \underline{3x^2 - 6x} \\ 4x - 11 \\ \underline{4x - 8} \\ -3 \end{array}$$

$(x^3 + x^2 - 2x - 11) \div (x - 2) = x^2 + 3x + 4 - \dfrac{3}{x - 2}$

Exercises 5.6

1. $\qquad 9x - 6y + 15 = 3(3x - 2y + 5)$
   Check: $3(3x - 2y + 5) = 9x - 6y + 15$

3. $\qquad 36x^2 - 9y^2 = 9(4x^2 - y^2)$
   $$= 9(2x + y)(2x - y)$$
   Check: $9(2x + y)(2x - y) = 9(4x^2 - y^2)$
   $$= 36x^2 - 9y^2$$

5. $9x^2 - 12xy + 4y^2 = (3x - 2y)^2$
   Check: $(3x - 2y)^2 = 9x^2 + 2(3x)(-2y) + (-2y)^2$
   $$= 9x^2 - 12xy + 4y^2$$

7. $\qquad x^2 + 3x - 28 = (x + 7)(x - 4)$
   Check: $(x + 7)(x - 4) = x^2 - 4x + 7x - 28$
   $$= x^2 + 3x - 28$$

9. $\qquad 5x^2 - 13x + 6 = (5x - 3)(x - 2)$
   Check: $(5x - 3)(x - 2) = 5x^2 - 10x - 3x + 6$
   $$= 5x^2 - 13x + 6$$

11. $ax - 3ay - 6by + 2bx$
    $$= a(x - 3y) + 2b(x - 3y)$$
    $$= (x - 3y)(a + 2b)$$
    Check: $(x - 3y)(a + 2b) = ax + 2bx - 3ay - 6by$
    $$= ax - 3ay - 6by + 2bx$$

13. $3x^4 - 12 = 3(x^4 - 4)$
    $$= 3(x^2 + 2)(x^2 - 2)$$

15. $2x^2 - 10x - 132 = 2(x^2 - 5x - 66)$
    $$= 2(x - 11)(x + 6)$$

17. $2x^2 - 11x + 12 = (2x - 3)(x - 4)$

19. $x^2 - 3xy - 70y^2 = (x - 10y)(x + 7y)$

21. $ax + 20 - 5a - 4x = ax - 5a - 4x + 20$
    $$= a(x - 5) - 4(x - 5)$$
    $$= (x - 5)(a - 4)$$

23. $45x - 5x^3 = 5x(9 - x^2)$
    $$= 5x(3 + x)(3 - x)$$

25. $5x^3y^3 - 10x^2y^3 + 5xy^3 = 5xy^3(x^2 - 2x + 1)$
    $$= 5xy^3(x - 1)^2$$

27. $27xyz^2 - 12xy = 3xy(9z^2 - 4)$
    $$= 3xy(3z + 2)(3z - 2)$$

29. $3x^2 + 6x - 105 = 3(x^2 + 2x - 35)$
    $$= 3(x + 7)(x - 5)$$

31. $5x^2 - 30x + 40 = 5(x^2 - 6x + 8)$
    $$= 5(x - 4)(x - 2)$$

33. $7x^2 - 2x^4 + 4 = -1(2x^4 - 7x^2 - 4)$
    $$= -1(2x^2 + 1)(x^2 - 4)$$
    $$= -1(2x^2 + 1)(x + 2)(x - 2)$$

35. $6x^2 - 3x + 2$ is prime.

37. $3x^2 - 15xy + 18y^2 = 3(x^2 - 5xy + 6y^2)$
    $$= 3(x - 3y)(x - 2y)$$

39. $36x^2 - 6xy - 6y^2 = 6(6x^2 - xy - y^2)$
$$= 6(3x + y)(2x - y)$$

41. $8x^2 + 28x - 16 = 4(2x^2 + 7x - 4)$
$$= 4(2x - 1)(x + 4)$$

43. $99x^4 - 11x^2 = 11x^2(9x^2 - 1)$
$$= 11x^2(3x + 1)(3x - 1)$$

45. $a^2x + 2a^2y - 3ax - 6ay$
$$= a(ax + 2ay - 3x - 6y)$$
$$= a[a(x + 2y) - 3(x + 2y)]$$
$$= a(x + 2y)(a - 3)$$

47. A polynomial that cannot be factored by the methods of Chapter 5 is called prime.

Cumulative Review Problems

49. x is a number
$$3x = 7x + 12$$
$$-4x = 12$$
$$x = -3$$
The number is -3.

Exercises 5.7

1. $x^2 - 3x - 10 = 0$
$$(x - 5)(x + 2) = 0$$

$x - 5 = 0 \qquad x + 2 = 0$
$x = 5 \qquad\qquad x = -2$

Check:

$(5)^2 - 3(5) - 10 \overset{?}{=} 0, \qquad (-2)^2 - 3(-2) - 10 \overset{?}{=} 0$
$25 - 15 - 10 \overset{?}{=} 0 \qquad\qquad 4 + 6 - 10 \overset{?}{=} 0$
$0 = 0 \qquad\qquad\qquad\qquad 0 = 0$

3. $x^2 + 13x + 40 = 0$
$$(x + 8)(x + 5) = 0$$

$x + 8 = 0 \qquad x + 5 = 0$
$x = -8 \qquad\quad x = -5$

Check:

$(-8)^2 + 13(-8) + 40 \overset{?}{=} 0 \qquad (-5)^2 + 13(-5) + 40 \overset{?}{=} 0$
$64 - 104 + 40 \overset{?}{=} 0 \qquad\qquad 25 - 65 + 40 \overset{?}{=} 0$
$0 = 0 \qquad\qquad\qquad\qquad 0 = 0$

5. $\qquad 2x^2 - 5x - 3 = 0$
$$2x^2 - 6x + x - 3 = 0$$
$$2x(x - 3) + 1(x - 3) = 0$$
$$(2x + 1)(x - 3) = 0$$

$2x + 1 = 0 \qquad x - 3 = 0$
$2x = -1 \qquad\quad x = 3$
$x = -\dfrac{1}{2}$

Roots: $-\dfrac{1}{2}, \; 3$

Check:

$2\left(-\dfrac{1}{2}\right)^2 - 5\left(-\dfrac{1}{2}\right) - 3 \overset{?}{=} 0 \qquad 2(3)^2 - 5(3) - 3 \overset{?}{=} 0$
$\dfrac{1}{2} + \dfrac{5}{2} - 3 \overset{?}{=} 0 \qquad\qquad 18 - 15 - 3 \overset{?}{=} 0$
$0 = 0 \qquad\qquad\qquad\qquad 0 = 0$

7. $\qquad 4x^2 - 11x - 3 = 0$
$$4x^2 - 12x + x - 3 = 0$$
$$4x(x - 3) + 1(x - 3) = 0$$
$$(4x + 1)(x - 3) = 0$$

$4x + 1 = 0 \qquad x - 3 = 0$
$4x = -1 \qquad\quad x = 3$
$x = -\dfrac{1}{4}$

Roots: $-\dfrac{1}{4}, \; 3$

7. continued

Check:

$$4\left(-\frac{1}{4}\right)^2 - 11\left(-\frac{1}{4}\right) - 3 \overset{?}{=} 0 \qquad 4(3)^2 - 11(3) - 3 \overset{?}{=} 0$$

$$\frac{1}{4} + \frac{11}{4} - 3 \overset{?}{=} 0 \qquad\qquad 36 - 33 - 3 \overset{?}{=} 0$$

$$0 = 0 \qquad\qquad\qquad\qquad 0 = 0$$

9. $\quad x^2 - 6x = 16$

$x^2 - 6x - 16 = 0$

$(x - 8)(x + 2) = 0$

$x - 8 = 0 \qquad x + 2 = 0$

$x = 8 \qquad\qquad x = -2$

Roots: 8, −2

Check:

$$(8)^2 - 6(8) \overset{?}{=} 16 \qquad (-2)^2 - 6(-2) \overset{?}{=} 16$$

$$64 - 48 \overset{?}{=} 16 \qquad\qquad 4 + 12 \overset{?}{=} 16$$

$$16 = 16 \qquad\qquad\qquad 16 = 16$$

11. $\quad 10 = 7x - x^2$

$x^2 - 7x + 10 = 0$

$(x - 5)(x - 2) = 0$

$x - 5 = 0 \qquad x - 2 = 0$

$x = 5 \qquad\qquad x = 2$

Roots: 5, 2

Check:

$$10 \overset{?}{=} 7(5) - (5)^2 \qquad 10 \overset{?}{=} 7(2) - (2)^2$$

$$10 \overset{?}{=} 35 - 25 \qquad\qquad 10 \overset{?}{=} 14 - 4$$

$$10 = 10 \qquad\qquad\qquad 10 = 10$$

13. $\quad 6x^2 - 13x = -6$

$6x^2 - 13x + 6 = 0$

$(3x - 2)(2x - 3) = 0$

13. continued

$3x - 2 = 0 \qquad\qquad 2x - 3 = 0$

$3x = 2 \qquad\qquad\qquad 2x = 3$

$x = \dfrac{2}{3} \qquad\qquad\qquad x = \dfrac{3}{2}$

Roots: $\dfrac{2}{3}, \dfrac{3}{2}$

Check:

$$6\left(\frac{2}{3}\right)^2 - 13\left(\frac{2}{3}\right) \overset{?}{=} -6 \qquad 6\left(\frac{3}{2}\right)^2 - 13\left(\frac{3}{2}\right) \overset{?}{=} -6$$

$$\frac{8}{3} - \frac{26}{3} \overset{?}{=} -6 \qquad\qquad \frac{27}{2} - \frac{39}{2} \overset{?}{=} -6$$

$$-6 = -6 \qquad\qquad\qquad -6 = -6$$

15. $\quad x^2 - 8x = 0$

$x(x - 8) = 0$

$x = 0 \qquad\qquad x - 8 = 0$

$\qquad\qquad\qquad x = 8$

Roots: 0, 8

Check:

$$(0)^2 - 8(0) \overset{?}{=} 0 \qquad (8)^2 - 8(8) \overset{?}{=} 0$$

$$0 - 0 \overset{?}{=} 0 \qquad\qquad 64 - 64 \overset{?}{=} 0$$

$$0 = 0 \qquad\qquad\qquad 0 = 0$$

17. $\quad 3x^2 = 2x$

$3x^2 - 2x = 0$

$x(3x - 2) = 0$

$x = 0 \qquad\qquad 3x - 2 = 0$

$\qquad\qquad\qquad 3x = 2$

$\qquad\qquad\qquad x = \dfrac{2}{3}$

Roots: $0, \dfrac{2}{3}$

Check:

$$3(0)^2 \overset{?}{=} 2(0) \qquad 3\left(\frac{2}{3}\right)^2 \overset{?}{=} 2\left(\frac{2}{3}\right)$$

$$0 = 0 \qquad\qquad\qquad \frac{4}{3} = \frac{4}{3}$$

19. $5x^2 + 3x = 8x$

$5x^2 - 5x = 0$

$5x(x-1) = 0$

$5x = 0 \qquad x - 1 = 0$

$x = 0 \qquad\quad x = 1$

Roots: 0, 1

Check:

$5(0)^2 + 3(0) \overset{?}{=} 8(0) \qquad 5(1)^2 + 3(1) \overset{?}{=} 8(1)$

$0 + 0 \overset{?}{=} 0 \qquad\qquad 5 + 3 \overset{?}{=} 8$

$0 = 0 \qquad\qquad\qquad 8 = 8$

21. $\quad x(9+x) = 4(2x+5)$

$9x + x^2 = 8x + 20$

$x^2 + x - 20 = 0$

$(x+5)(x-4) = 0$

$x + 5 = 0 \qquad x - 4 = 0$

$x = -5 \qquad\quad x = 4$

Roots: -5, 4

Check:

$(-5)(9-5) \overset{?}{=} 4[2(-5)+5] \qquad 4(9+4) \overset{?}{=} 4(2\cdot4+5)$

$(-5)(4) \overset{?}{=} 4(-5) \qquad\qquad 4(13) \overset{?}{=} 4(13)$

$-20 = -20 \qquad\qquad\qquad 52 = 52$

23. $x^2 - 12x + 36 = 0$

$(x-6)(x-6) = 0$

$x - 6 = 0$

$x = 6$

Root: 6

Check:

$(6)^2 - 12(6) + 36 \overset{?}{=} 0$

$36 - 72 + 36 \overset{?}{=} 0$

$0 = 0$

25. $\qquad 4x^2 - 3x + 1 = -7x$

$4x^2 + 4x + 1 = 0$

$(2x+1)(2x+1) = 0$

$2x + 1 = 0$

$2x = -1$

$x = -\dfrac{1}{2}$

Root: $-\dfrac{1}{2}$

Check:

$4\left(-\dfrac{1}{2}\right)^2 - 3\left(-\dfrac{1}{2}\right) + 1 \overset{?}{=} -7\left(-\dfrac{1}{2}\right)$

$1 + \dfrac{3}{2} + 1 \overset{?}{=} \dfrac{7}{2}$

$\dfrac{7}{2} = \dfrac{7}{2}$

27. $\dfrac{x^2}{2} - 8 + x = -8$

$x^2 - 16 + 2x = -16$

$x^2 + 2x = 0$

$x(x+2) = 0$

$x = 0 \qquad x + 2 = 0$

$x = -2$

Roots: 0, -2

Check:

$\dfrac{(0)^2}{2} - 8 + (0) \overset{?}{=} -8 \qquad \dfrac{(-2)^2}{2} - 8 + (-2) \overset{?}{=} -8$

$0 - 8 + 0 \overset{?}{=} -8 \qquad\qquad 2 - 8 - 2 \overset{?}{=} -8$

$-8 = -8 \qquad\qquad\qquad -8 = -8$

29. $\quad \dfrac{x^2 - 3x}{2} = 27$

$x^2 - 3x = 54$

$x^2 - 3x - 54 = 0$

$(x-9)(x+6) = 0$

29. continued

$$x - 9 = 0 \qquad x + 6 = 0$$
$$x = 9 \qquad\qquad x = -6$$

Roots: $9, \ -6$

Check:

$$\frac{(9)^2 - 3(9)}{2} \overset{?}{=} 27 \qquad\qquad \frac{(-6)^2 - 3(-6)}{2} \overset{?}{=} 27$$

$$\frac{81 - 27}{2} \overset{?}{=} 27 \qquad\qquad \frac{36 + 18}{2} \overset{?}{=} 27$$

$$27 = 27 \qquad\qquad\qquad 27 = 27$$

31.
$$\frac{2x^2 + x}{3} = 12$$

$$2x^2 + x = 36$$

$$2x^2 + x - 36 = 0$$

$$(2x + 9)(x - 4) = 0$$

$$2x + 9 = 0 \qquad\qquad x - 4 = 0$$

$$x = -\frac{9}{2} \qquad\qquad x = 4$$

Roots: $-\dfrac{9}{2}, \ 4$

Check:

$$\frac{2\left(-\dfrac{9}{2}\right)^2 + \left(-\dfrac{9}{2}\right)}{3} \overset{?}{=} 12 \qquad\qquad \frac{2(4)^2 + (4)}{3} \overset{?}{=} 12$$

$$\frac{\dfrac{81}{2} - \dfrac{9}{2}}{3} \overset{?}{=} 12 \qquad\qquad\qquad \frac{32 + 4}{3} \overset{?}{=} 12$$

$$12 = 12 \qquad\qquad\qquad\qquad 12 = 12$$

To Think About

33. An equation in the form
$ax^2 + bx = 0$ can always
be solved by factoring out x.

35. $x =$ length
$$\frac{1}{2}x + 3 = \text{ width}$$

$$x\left(\frac{1}{2}x + 3\right) = 140$$

$$\frac{x^2}{2} + 3x = 140$$

$$x^2 + 6x = 280$$

$$x^2 + 6x - 280 = 0$$

$$(x + 20)(x - 14) = 0$$

$$x + 20 = 0 \qquad\qquad x - 14 = 0$$
$$x = -20 \qquad\qquad x = 14$$

Width not negative $\dfrac{1}{2}x + 3 = 10$

The length is 14 m and the width is 10 m.

37. $x =$ 1st number
$x + 10 =$ 2nd number

$$x(x + 10) = 56$$

$$x^2 + 10x = 56$$

$$x^2 + 10x - 56 = 0$$

$$(x + 14)(x - 4) = 0$$

$$x + 14 = 0 \qquad\qquad x - 4 = 0$$
$$x = -14 \qquad\qquad x = 4$$

Can't be negative $x + 10 = 14$

The numbers are 4 and 14.

39. $s = -5t^2 + vt + h$
$$v = 13 \qquad h = 6$$
$$s = -5t^2 + 13t + 6$$

When the ball hits the ground $s = 0$.

$$0 = -5t^2 + 13t + 6$$

$$0 = 5t^2 - 13t - 6$$

$$0 = (5t + 2)(t - 3)$$

39. continued

$5t + 2 = 0 \qquad\qquad t - 3 = 0$

$t = -\dfrac{2}{5} \qquad\qquad t = 3$

Time can't be negative

The ball will hit the ground at $t = 3$ sec.

$s = -5(2)^2 + 13(2) + 6$

$\quad = -20 + 26 + 6$

$\quad = 12$

After 2 seconds the ball is 12 m from the ground.

41. $C = 2x^2 - 7x$

$15 = 2x^2 - 7x$

$2x^2 - 7x - 15 = 0$

$(2x + 3)(x - 5) = 0$

$2x + 3 = 0 \qquad\qquad x - 5 = 0$

$x = -\dfrac{3}{2} \qquad\qquad x = 5$

Can't be negative

5 additional stereos were produced.

Cumulative Review Problems

43. $(2x^2 y^3)(-6xy^4) = -12x^3 y^7$

45. $\dfrac{21a^5 b^{10}}{-14ab^{12}} = -\dfrac{3a^4}{2b^2}$

Putting Your Skills to Work

1. $\qquad w = $ width

$\left(2\dfrac{1}{2}\right)w = $ length

Overall width $= w + 10$

Overall length $= \left(2\dfrac{1}{2}\right)w + 10$

Area of deck $=$ Overall area $-$ area of pool.

1. continued

$1150 = (w + 10)\left(\dfrac{5}{2} w + 10\right) - \left(\dfrac{5}{2} w\right)w$

$1150 = \dfrac{5}{2} w^2 + 35w + 100 - \dfrac{5}{2} w^2$

$1150 = 35w + 100$

$1050 = 35w$

$30 = w$

$\left(2\dfrac{1}{2}\right)w = 75$

Length is 75 feet, width is 30 ft.

3. Volume $=$ Area $\times$ depth

$= (1150)\left(\dfrac{1}{4}\right)$

$= 287.5 \ \text{ft}^3$

$= \dfrac{287.5}{27} \ \text{yd}^3$

$= 10.64815 \ \text{yd}^3$

Cost $= 125(10.64815) = \$1331.02$

Chapter 5 - Review Problems

1. $15x^3 y - 9x^2 y^2 = 3x^2 y(5x - 3y)$

3. $7x^2 y - 14xy^2 - 21x^3 y^3 = 7xy(x - 2y - 3x^2 y^2)$

5. $27x^3 - 9x^2 = 9x^2(3x - 1)$

7. $2a(a + 3b) - 5(a + 3b) = (a + 3b)(2a - 5)$

9. $3ax - 7a - 6x + 14 = a(3x - 7) - 2(3x - 7)$
$\qquad\qquad\qquad\qquad = (3x - 7)(a - 2)$

11. $x^2 y + 3y - 2x^2 - 6 = y(x^2 + 3) - 2(x^2 + 3)$
$\qquad\qquad\qquad\qquad = (x^2 + 3)(y - 2)$

13. $15x^2 - 3x + 10x - 2 = 3x(5x - 1) + 2(5x - 1)$
$\qquad\qquad\qquad\qquad = (5x - 1)(3x + 2)$

15. $x^2 - 3x - 18 = (x - 6)(x + 3)$

17. $x^2 + 14x + 48 = (x + 6)(x + 8)$

90

19. $x^4 + 13x^2 + 42 = (x^2 + 6)(x^2 + 7)$

21. $5x^2 + 20x + 15 = 5(x^2 + 4x + 3)$
$\qquad = 5(x + 3)(x + 1)$

23. $2x^2 - 28x + 96 = 2(x^2 - 14x + 48)$
$\qquad = 2(x - 8)(x - 6)$

25. $4x^2 + 7x - 15 = (4x - 5)(x + 3)$

27. $15x^2 + 7x - 4 = (3x - 1)(5x + 4)$

29. $2x^2 - x - 3 = (2x - 3)(x + 1)$

31. $20x^2 + 48x - 5 = (10x - 1)(2x + 5)$

33. $4a^2 - 11a - 3 = (4a + 1)(a - 3)$

35. $6x^2 + 4x - 10 = 2(3x^2 + 2x - 5)$
$\qquad = 2(3x + 5)(x - 1)$

37. $4x^2 - 26x + 30 = 2(2x^2 - 13x + 15)$
$\qquad = 2(2x - 3)(x - 5)$

39. $10x^2 - 22x + 12 = 2(5x^2 - 11x + 6)$
$\qquad = 2(5x - 6)(x - 1)$

41. $12x^2 + 5x - 3 = (4x + 3)(3x - 1)$

43. $6x^2 - 19xy + 10y^2 = (2x - 5y)(3x - 2y)$

45. $3x^4 - 5x^2 - 8 = (3x^2 - 8)(x^2 + 1)$

47. $49x^2 - y^2 = (7x)^2 - (y)^2$
$\qquad = (7x + y)(7x - y)$

49. $9x^2 - 12x + 4 = (3x - 2)^2$

51. $25x^2 - 36 = (5x)^2 - (6)^2$
$\qquad = (5x + 6)(5x - 6)$

53. $1 - 49x^2 = (1)^2 - (7x)^2$
$\qquad = (1 + 7x)(1 - 7x)$

55. $36x^2 + 12x + 1 = (6x + 1)^2$

57. $16x^2 - 24xy + 9y^2 = (4x - 3y)^2$

59. $2x^2 - 18 = 2(x^2 - 9)$
$\qquad = 2(x + 3)(x - 3)$

61. $8x^2 + 40x + 50 = 2(4x^2 + 20x + 25)$
$\qquad = 2(2x + 5)^2$

63. $4x^2 - 9y^2 = (2x)^2 - (3y)^2$
$\qquad = (2x + 3y)(2x - 3y)$

65. $x^2 - 9x + 18 = (x - 6)(x - 3)$

67. $6x^2 + x - 7 = (6x + 7)(x - 1)$

69. $12x + 16 = 4(3x + 4)$

71. $50x^3y^2 + 30x^2y^2 - 10x^2y^2 = 50x^3y^2 + 20x^2y^2$
$\qquad = 10x^2y^2(5x + 2)$

73. $x^3 - 16x^2 + 64x = x(x^2 - 16x + 64)$
$\qquad = x(x - 8)^2$

75. $3x^2 - 18x + 27 = 3(x^2 - 6x + 9)$
$\qquad = 3(x - 3)^2$

77. $7x^2 + 3x - 10 = (7x + 10)(x - 1)$

79. $9x^3y - 4xy^3 = xy(9x^2 - 4y^2)$
$\qquad = xy(3x + 2y)(3x - 2y)$

81. $12a^2 + 14ab - 10b^2 = 2(6a^2 + 7ab - 5b^2)$
$\qquad = 2(3a + 5b)(2a - b)$

83. $7a - 7 - ab + b = 7(a - 1) - b(a - 1)$
$\qquad = (a - 1)(7 - b)$

85. $2x - 1 + 2bx - b = (2x - 1) + b(2x - 1)$
$\qquad = (2x - 1)(1 + b)$

87. $2a^2x - 15ax + 7x = x(2a^2 - 15a + 7)$
$\qquad = x(2a - 1)(a - 7)$

89. $x^4 - 81y^{12} = \left(x^2\right)^2 - \left(9y^6\right)^2$

$$= \left(x^2 + 9y^6\right)\left(x^2 - 9y^6\right)$$

$$= \left(x^2 + 9y^6\right)\left(x + 3y^3\right)\left(x - 3y^3\right)$$

91. $28yz - 16xyz + x^2yz = yz\left(28 - 16x + x^2\right)$

$$= yz(14 - x)(2 - x)$$

93. $16w^2 - 2w - 5 = (2w + 1)(8w - 5)$

95. $4y^3 + 10y^2 - 6y = 2y\left(2y^2 + 5y - 3\right)$

$$= 2y(2y - 1)(y + 3)$$

97. $8y^{10} - 16y^8 = 8y^8\left(y^2 - 2\right)$

99. $x^2 + 13x + 54$ is prime.

101. $8y^5 + 4y^3 - 60y = 4y\left(2y^4 + y^2 - 15\right)$

$$= 4y\left(2y^2 - 5\right)\left(y^2 + 3\right)$$

103. $16x^4y^2 - 56x^2y + 49 = \left(4x^2y - 7\right)^2$

105. $2ax + 5a - 10b - 4bx = a(2x + 5) - 2b(2x + 5)$

$$= (2x + 5)(a - 2b)$$

107. $x^2 - 3x - 18 = 0$

$(x - 6)(x + 3) = 0$

$x - 6 = 0 \qquad x + 3 = 0$

$x = 6 \qquad\quad x = -3$

Roots: $-3, \ 6$

109. $\qquad 5x^2 = 2x - 7x^2$

$12x^2 - 2x = 0$

$2x(6x - 1) = 0$

$2x = 0 \qquad 6x - 1 = 0$

$x = 0 \qquad\quad x = \dfrac{1}{6}$

Roots: $0, \ \dfrac{1}{6}$

111. $2x^2 + 9x - 5 = 0$

$(2x - 1)(x + 5) = 0$

$2x - 1 = 0 \qquad\quad x + 5 = 0$

$x = \dfrac{1}{2} \qquad\qquad x = -5$

Roots: $\dfrac{1}{2}, \ -5$

113. $x^2 + 14x + 45 = 0$

$(x + 9)(x + 5) = 0$

$x + 9 = 0 \qquad\quad x + 5 = 0$

$x = -9 \qquad\qquad x = -5$

Roots: $-9, \ -5$

115. $\quad 3x^2 + 6x = 2x^2 - 9$

$x^2 + 6x + 9 = 0$

$(x + 3)^2 = 0$

$x + 3 = 0$

$x = -3$

Roots: -3

117. $5x^2 - 11x + 2 = 0$

$(5x - 1)(x - 2) = 0$

$5x - 1 = 0 \qquad\quad x - 2 = 0$

$x = \dfrac{1}{5} \qquad\qquad x = 2$

Roots: $\dfrac{1}{5}, \ 2$

119. $\qquad x = \text{width}$

$2x + 1 = \text{length}$

$A = \text{length} \times \text{width}$

$105 = x(2x + 1)$

$105 = 2x^2 + x$

$0 = 2x^2 + x - 105$

$0 = (2x + 15)(x - 7)$

119. continued

$$2x + 15 = 0 \quad \text{or} \quad x - 7 = 0$$
$$x = -\frac{15}{2} \qquad x = 7$$
$$2x + 1 = 15$$

Length can't be negative so width
= 7 feet and length = 15 feet.

121.
$$480 = -5x^2 + 100x$$
$$5x^2 - 100x + 480 = 0$$
$$5(x^2 - 20x + 96) = 0$$

$$x - 12 = 0 \quad \text{or} \quad x - 8 = 0$$
$$x = 12 \qquad x = 8$$

The current is 12 amperes or 8 amperes.

Putting Your Skills to Work

Use $P = -\frac{7}{400}t^2 + 7t - 400$.

1. $t = 100$

$$P = -\frac{7}{400}(100)^2 + 7(100) - 400 = 125$$
Profit is $125.00

3. $a = -\frac{7}{400}$, $b = 7$

Maximum: $t = \dfrac{-7}{2\left(-\dfrac{7}{400}\right)} = 200$

Need to sell 200 items.

5. $t = 199$:

$$P = -\frac{7}{400}(199)^2 + 7(199) - 400 = 299.98$$
$t = 201$:
$$P = -\frac{7}{400}(201)^2 + 7(201) - 400 = 299.98$$

The maximum profit from problem 4 is $300.00.
If the items sold is greater than or less than 200,
the profit is less than $300.00.

Use $P = -n^2 + 3000n - 2,000,000$ in Exercises 7-11.

7. $n = 500$:
$$P = -(500)^2 + 3000(500) - 2,000,000$$
$$= -750,000$$
Loss is $750,000.

9. Maximum profit occurs at
$$n = \frac{-b}{2a} = \frac{-3000}{2(-1)} = 1500$$

Sell 1500 cars to maximize profit.

11. $n = 0$:
$$P = -(0)^2 + 3000(0) - 2,000,000$$
$$= -2,000,000$$
Loss is $2,000,000.

Test - Chapter 5

1. $x^2 + 12x - 28 = (x + 14)(x - 2)$

3. $10x^2 + 27x + 5 = (5x + 1)(2x + 5)$

5. $7x - 9x^2 + 14xy = x(7 - 9x + 14y)$

7. $6x^3 - 20x^2 + 16x = 2x(3x^2 - 10x + 8)$
$$= 2x(3x - 4)(x - 2)$$

9. $100x^4 - 16y^4 = 4(25x^4 - 4y^4)$
$$= 4\left[(5x^2)^2 - (2y^2)^2\right]$$
$$= 4(5x^2 + 2y^2)(5x^2 - 2y^2)$$

11. $7x^2 - 42x = 7x(x - 6)$

13. $3x^2 + 5x + 1$ is prime.

15. $16x^2 - 1 = (4x)^2 - (1)^2$
$$= (4x + 1)(4x - 1)$$

17. $2ax + 6a - 5x - 15 = 2a(x + 3) - 5(x + 3)$
$$= (x + 3)(2a - 5)$$

19. $3x^2 - 3x - 90 = 3(x^2 - x - 30)$

$\qquad = 3(x - 6)(x + 5)$

21. $x^2 + 14x + 45 = 0$

$(x + 9)(x + 5) = 0$

$x + 9 = 0 \qquad\qquad x + 5 = 0$

$x = -9 \qquad\qquad x = -5$

Roots: $-9, \ -5$

23. $2x^2 + x - 10 = 0$

$(2x + 5)(x - 2) = 0$

$2x + 5 = 0 \qquad\qquad x - 2 = 0$

$x = -\dfrac{5}{2} \qquad\qquad x = 2$

Roots: $-\dfrac{5}{2}, \ 2$

Cumulative Test - Chapters 0 - 5

1. $\dfrac{72}{480} = 0.15 = 15\%$

3. $-3.2 - 6.4 + 0.24 - 1.8 + 0.8$

$= -11.4 + 1.04$

$= -10.36$

5. $-2^6 = -64$

7. $(2x^2 - 6x + 1)(x - 3) = 2x^3 - 6x^2 + x - 6x^2 + 18x - 3$

$\qquad\qquad\qquad = 2x^3 - 12x^2 + 19x - 3$

9. $3x - (7 - 5x) = 3(4x - 5)$

$3x - 7 + 5x = 12x - 15$

$-7 + 8x = 12x - 15$

$8 = 4x$

$2 = x$

11. $\qquad s = \dfrac{1}{2}(2a + 3t)$

$2s = 2a + 3t$

$2s - 2a = 3t$

$\dfrac{2s - 2a}{3} = t$

13. $6x^2 - 5x - 1 = (6x + 1)(x - 1)$

15. $121x^2 - 64y^2 = (11x)^2 - (8y)^2$

$\qquad\qquad = (11x + 8y)(11x - 8y)$

17. $x^2 + 1 = $ prime.

19. $81x^4 - 16b^4 = (9x^2)^2 - (4b^2)^2$

$\qquad\qquad = (9x^2 + 4b^2)(9x^2 - 4b^2)$

$\qquad\qquad = (9x^2 + 4b^2)(3x + 2b)(3x - 2b)$

21. $x^4 + 8x^2 + 15 = (x^2 + 5)(x^2 + 3)$

23. $3x^2 - 11x + 10 = 0$

$(3x - 5)(x - 2) = 0$

$3x - 5 = 0 \qquad\qquad x - 2 = 0$

$x = \dfrac{5}{3} \qquad\qquad x = 2$

Roots: $\dfrac{5}{3}, \ 2$

Practice Quiz: Sections 5.1 - 5.3

1. Remove the greatest common factor.

$14x^3y - 21x^2y^2 + 28xy^3$

2. Remove the greatest common factor.

$a(2b - 5) + 3b(2b - 5)$

3. Factor: $x^2 + 3x - 2x - 6$

4. Factor: $x^2 + 9x + 18$

5. Factor: $x^2 + 2x - 35$

Practice Quiz: Sections 5.4 - 5.7

1. Factor: $6x^2 - 17x - 14$

2. Factor: $25x^2 - 49y^2$

3. Factor: $16a^2 - 56ab + 49b^2$

4. Factor: $16x^4 - y^4$

5. Solve: $2x^2 + 7x - 15 = 0$

Answers to Practice Quiz

Sections 5.1 - 5.3

1. $7xy\left(2x^2 - 3xy + 4y^2\right)$

2. $(2b - 5)(a + 3b)$

3. $(x - 2)(x + 3)$

4. $(x + 3)(x + 6)$

5. $(x - 5)(x + 7)$

Answers to Practice Quiz

Sections 5.4 - 5.7

1. $(3x - 2)(2x + 7)$

2. $(5x + 7y)(5x - 7y)$

3. $(4a - 7b)^2$

4. $\left(4x^2 + y^2\right)(2x + y)(2x - y)$

5. Roots: $-5, \dfrac{3}{2}$

1. $\dfrac{6a-4b}{2b-3a} = \dfrac{-2(-3a+2b)}{2b-3a} = \dfrac{-2(2b-3a)}{2b-3a}$

    $= \dfrac{-2(2b-3a)}{2b-3a} = -2$

3. $\dfrac{a^2b+2ab^2}{2a^3+3a^2b-2ab^2} = \dfrac{ab(a+2b)}{a(2a^2+3ab-2b^2)}$

    $= \dfrac{ab(a+2b)}{a(2a-b)(a+2b)} = \dfrac{ab(a+2b)}{a(2a-b)(a+2b)}$

    $= \dfrac{b}{2a-b}$

5. $\dfrac{x^2-6x+9}{x^2-x-6} \div \dfrac{x^2+2x-15}{x^2+2x}$

    $= \dfrac{x^2-6x+9}{x^2-x-6} \cdot \dfrac{x^2+2x}{x^2+2x-15}$

    $= \dfrac{(x-3)(x-3)}{(x-3)(x+2)} \cdot \dfrac{x(x+2)}{(x+5)(x-3)}$

    $= \dfrac{(x-3)(x-3)}{(x-3)(x+2)} \cdot \dfrac{x(x+2)}{(x+5)(x-3)} = \dfrac{x}{x+5}$

7. $\dfrac{xy+3y}{x^2-x} \div \dfrac{x+3}{x} = \dfrac{xy+3y}{x^2-x} \cdot \dfrac{x}{x+3}$

    $= \dfrac{y(x+3)}{x(x-1)} \cdot \dfrac{x}{x+3} = \dfrac{y}{x-1}$

9. $\dfrac{2y-1}{2y^2+y-3} - \dfrac{2}{y-1} = \dfrac{2y-1}{(2y+3)(y-1)} - \dfrac{2}{y-1}$

    $= \dfrac{2y-1}{(2y+3)(y-1)} - \dfrac{2(2y+3)}{(y-1)(2y+3)}$

    $= \dfrac{2y-1-2(2y+3)}{(2y+3)(y-1)} = \dfrac{2y-1-4y-6}{(2y+3)(y-1)}$

    $= \dfrac{-2y-7}{(2y+3)(y-1)}$

11. $\dfrac{x+3}{x^2-6x+9} + \dfrac{2x+3}{3x^2-9x}$

    $= \dfrac{x+3}{(x-3)(x-3)} + \dfrac{2x+3}{3x(x-3)}$

    $= \dfrac{3x(x+3)}{3x(x-3)(x-3)} + \dfrac{(2x+3)(x-3)}{3x(x-3)(x-3)}$

    $= \dfrac{3x^2+9x+2x^2-3x-9}{3x(x-3)(x-3)} = \dfrac{5x^2+6x-9}{3x(x-3)(x-3)}$

    $= \dfrac{5x^2+6x-9}{3x(x-3)^2}$

13. $\dfrac{\frac{a}{a+1} - \frac{2}{a}}{3a} = \dfrac{\frac{a(a)}{a(a+1)} - \frac{2(a+1)}{a(a+1)}}{3a}$

    $= \dfrac{\frac{a^2-2(a+1)}{a(a+1)}}{3a} = \dfrac{\frac{a^2-2a-2}{a(a+1)}}{3a}$

    $= \dfrac{a^2-2a-2}{a(a+1)} \cdot \dfrac{1}{3a} = \dfrac{a^2-2a-2}{3a^2(a+1)}$

15. $\dfrac{5}{2x} = 2 - \dfrac{2x}{x+1}$

    $(2x)(x+1)\left(\dfrac{5}{2x}\right) = 2x(x+1)(2) - (2x)(x+1)\left(\dfrac{2x}{x+1}\right)$

    $(x+1)(5) = (2x^2+2x)(2) - (2x)(2x)$

    $5x+5 = 4x^2+4x-4x^2$

    $5x+5 = 4x$

    $5 = -x$

    $-5 = x$

17. $\dfrac{5}{x} = \dfrac{7}{13}$

    $65 = 7x$

    $\dfrac{65}{7} = x$

    $x = 9.3$

Exercises 6.1

1. $\dfrac{6x-3y}{2x-y} = \dfrac{3(2x-y)}{2x-y} = 3$

3. $\dfrac{x^2}{3x^2-x^3} = \dfrac{x^2}{x^2(3-x)} = \dfrac{1}{3-x}$

5. $\dfrac{2x-8}{x^2-8x+16} = \dfrac{2(x-4)}{(x-4)(x-4)} = \dfrac{2}{x-4}$

7. $\dfrac{x^2-9y^2}{x+3y} = \dfrac{(x-3y)(x+3y)}{x+3y} = x-3y$

9. $\dfrac{6x^2}{2x(x-3y)} = \dfrac{3x}{x-3y}$

11. $\dfrac{x^2+x-2}{x^2-x} = \dfrac{(x+2)(x-1)}{x(x-1)} = \dfrac{x+2}{x}$

13. $\dfrac{x^2+2x-3}{4x^2-5x+1} = \dfrac{(x+3)(x-1)}{(x-1)(4x-1)} = \dfrac{x+3}{4x-1}$

15. $\dfrac{x^3-8x^2+16x}{x^2+2x-24} = \dfrac{x(x^2-8x+16)}{(x+6)(x-4)} = \dfrac{x(x-4)(x-4)}{(x+6)(x-4)}$

$= \dfrac{x(x-4)}{x+6}$

17. $\dfrac{x^2+14x+45}{2x^2+17x-9} = \dfrac{(x+9)(x+5)}{(x+9)(2x-1)} = \dfrac{x+5}{2x-1}$

19. $\dfrac{3x^2-8x+5}{4x^2-5x+1} = \dfrac{(3x-5)(x-1)}{(4x-1)(x-1)} = \dfrac{3x-5}{4x-1}$

21. $\dfrac{2x^2-5x-12}{2x^2-x-6} = \dfrac{(2x+3)(x-4)}{(2x+3)(x-2)} = \dfrac{x-4}{x-2}$

23. $\dfrac{2x^3+11x^2+5x}{2x^3-5x^2-3x} = \dfrac{x(2x^2+11x+5)}{x(2x^2-5x-3)}$

$= \dfrac{x(2x+1)(x+5)}{x(2x+1)(x-3)} = \dfrac{x+5}{x-3}$

25. $\dfrac{6-3x}{2x-4} = \dfrac{-3x+6}{2x-4} = \dfrac{-3(x-2)}{2(x-2)} = \dfrac{-3}{2}$

27. $\dfrac{2x^2-7x-15}{25-x^2} = \dfrac{(2x+3)(x-5)}{(5-x)(5+x)}$

$= \dfrac{-2x-3}{x+5} = \dfrac{-2x-3}{5+x}$

29. $\dfrac{(4x+5)^2}{8x^2+6x-5} = \dfrac{(4x+5)^2}{(4x+5)(2x-1)}$

$= \dfrac{(4x+5)(4x+5)}{(4x+5)(2x-1)} = \dfrac{4x+5}{2x-1}$

31. $\dfrac{2y^2-5y-12}{8+2y-y^2} = \dfrac{(2y+3)(y-4)}{(4-y)(2+y)} = \dfrac{-2y-3}{2+y}$

33. $\dfrac{a^2+2ab-3b^2}{2a^2+5ab-3b^2} = \dfrac{(a+3b)(a-b)}{(2a-b)(a+3b)} = \dfrac{a-b}{2a-b}$

35. $\dfrac{16x^2-25y^2}{4x^2+3xy-10y^2} = \dfrac{(4x-5y)(4x+5y)}{(4x-5y)(x+2y)}$

$= \dfrac{4x+5y}{x+2y}$

37. $\dfrac{6x^4-9x^3-6x^2}{12x^3+42x^2+18x} = \dfrac{3x^2(2x^2-3x-2)}{6x(2x^2+7x+3)}$

$= \dfrac{3x^2(2x+1)(x-2)}{6x(2x+1)(x+3)}$

$= \dfrac{x(x-2)}{2(x+3)}$

Cumulative Review Problems

39. $(3x-7)^2 = (3x)^2 - 2(3x)(7) + (7)^2$

$= 9x^2 - 42x + 49$

41. $(2x+3)(x-4)(x-2)$

$= (2x+3)(x^2-6x+8)$

$= 2x^3 - 12x^2 + 16x + 3x^2 - 18x + 24$

$= 2x^3 - 9x^2 - 2x + 24$

1. We always first try to simplify.

3. $\dfrac{x+3}{x+7} \cdot \dfrac{x^2+3x-10}{x^2+x-6} = \dfrac{x+3}{x+7} \cdot \dfrac{(x+5)(x-2)}{(x+3)(x-2)}$

$= \dfrac{x+5}{x+7}$

5. $\dfrac{x^2+2x}{6x} \cdot \dfrac{3x^2}{x^2-4} = \dfrac{x(x+2)}{6x} \cdot \dfrac{3x^2}{(x+2)(x-2)}$

$= \dfrac{x^2}{2(x-2)}$

7. $\dfrac{x^2-x-6}{x^2-2x-8} \cdot \dfrac{x^2+7x+12}{x^2-9}$

$= \dfrac{(x-3)(x+2)}{(x-4)(x+2)} \cdot \dfrac{(x+4)(x+3)}{(x-3)(x+3)} = \dfrac{x+4}{x-4}$

9. $\dfrac{x^2+7x-8}{2x^2-18} \cdot \dfrac{2x^2+20x+42}{7x^2-7x}$

$= \dfrac{(x+8)(x-1)}{2(x^2-9)} \cdot \dfrac{2(x^2+10x+21)}{7x(x-1)}$

$= \dfrac{(x+8)(x-1)}{2(x+3)(x-3)} \cdot \dfrac{2(x+7)(x+3)}{7x(x-1)} = \dfrac{(x+8)(x+7)}{7x(x-3)}$

11. $\dfrac{5x^2+6x+1}{x^2+5x+6} \div (5x+1) = \dfrac{(5x+1)(x+1)}{(x+3)(x+2)} \cdot \dfrac{1}{(5x+1)}$

$= \dfrac{x+1}{(x+3)(x+2)}$

13. $\dfrac{x^2+2xy+y^2}{x^2-2xy+y^2} \div \dfrac{3x+3y}{4x-4y}$

$= \dfrac{(x+y)(x+y)}{(x-y)(x-y)} \div \dfrac{3(x+y)}{4(x-y)}$

$= \dfrac{(x+y)(x+y)}{(x-y)(x-y)} \cdot \dfrac{4(x-y)}{3(x+y)} = \dfrac{4(x+y)}{3(x-y)}$

15. $\dfrac{xy-y^2}{x^2+2x+1} \div \dfrac{2x^2+xy-3y^2}{2x^2+5xy+3y^2}$

$= \dfrac{y(x-y)}{(x+1)(x+1)} \div \dfrac{(2x+3y)(x-y)}{(2x+3y)(x+y)}$

$= \dfrac{y(x-y)}{(x+1)(x+1)} \cdot \dfrac{(2x+3y)(x+y)}{(2x+3y)(x-y)} = \dfrac{y(x+y)}{(x+1)^2}$

17. $\dfrac{x^2+5x-14}{(x-5)} \div \dfrac{x^2+12x+35}{15-3x}$

$= \dfrac{(x+7)(x-2)}{(x-5)} \div \dfrac{(x+7)(x+5)}{3(5-x)}$

$= \dfrac{(x+7)(x-2)}{(x-5)} \cdot \dfrac{3(5-x)}{(x+7)(x+5)} = \dfrac{-3(x-2)}{x+5}$

19. $\dfrac{(x+4)^2}{2x^2-7x-15} \div \dfrac{(x-5)^2}{(x+4)}$

$= \dfrac{(x+4)(x+4)}{(2x+3)(x-5)} \cdot \dfrac{(x-5)(x-5)}{(x+4)} = \dfrac{(x+4)(x-5)}{2x+3}$

21. $\dfrac{y^2-y-2}{y^2+4y+3} \cdot \dfrac{y^2+y-6}{y^2-4y+4}$

$= \dfrac{(y-2)(y+1)}{(y+3)(y+1)} \cdot \dfrac{(y+3)(y-2)}{(y-2)(y-2)} = \dfrac{1}{1} = 1$

23. $\dfrac{3x^2-7x+2}{3x^2+2x-1} \cdot \dfrac{2x^2-9x-5}{x^2+x-6} \cdot \dfrac{4x^2+11x-3}{x^2-11x+30}$

$= \dfrac{(3x-1)(x-2)}{(3x-1)(x+1)} \cdot \dfrac{(2x+1)(x-5)}{(x+3)(x-2)} \cdot \dfrac{(x+3)(4x-1)}{(x-6)(x-5)}$

$= \dfrac{(2x+1)(4x-1)}{(x+1)(x-6)}$

Cumulative Review Problems

25. $5x^2-7x+11 = 5x^2-x+2$

$-7x+11 = -x+2$

$11 = 6x+2$

$9 = 6x$

$\dfrac{9}{6} = x$

$\dfrac{3}{2} = x$

1. $\dfrac{7}{12x^3} + \dfrac{1}{12x^3} = \dfrac{7+1}{12x^3} = \dfrac{8}{12x^3} = \dfrac{2}{3x^3}$

3. $\dfrac{5}{x+2} + \dfrac{x+3}{2+x} = \dfrac{5+x+3}{x+2} = \dfrac{x+8}{x+2}$

5. $\dfrac{x}{x-4} - \dfrac{x+1}{x-4} = \dfrac{x-(x+1)}{x-4} = -\dfrac{1}{x-4}$

7. $\dfrac{8x+3}{5x+7} - \dfrac{6x+10}{5x+7} = \dfrac{8x+3-(6x+10)}{5x+7}$

$= \dfrac{2x-7}{5x+7}$

9. $\dfrac{3x-2}{5x-6} - \dfrac{x+5}{5x-6} = \dfrac{3x-2-(x+5)}{5x-6}$

$= \dfrac{2x-7}{5x-6}$

11. $\dfrac{12}{5a^2}, \dfrac{9}{a^3}$

$\left. 5a^2 \right\}$

$\left. a^3 \right\}$ LCD $= 5a^3$

13. $\dfrac{7}{15x^3y^5}, \dfrac{13}{40x^4y}$

$\left. 15x^3y^5 = 3 \cdot 5x^3y^5 \right\}$ LCD $= 2^3 \cdot 3 \cdot 5x^4y^5$

$\left. 40x^4y = 2^3 \cdot 5x^4y \right\}$ $= 120x^4y^5$

15. $\dfrac{13}{x^2-16}, \dfrac{7}{x-4}$

$x^2 - 16 = (x+4)(x-4)$

$\dfrac{x-4 = x-4}{\text{LCD} = (x+4)(x-4)} = x^2 - 16$

17. $\dfrac{3}{16x^2 - 8x + 1}, \dfrac{6}{4x^2 + 7x - 2}$

$16x^2 - 8x + 1 = (4x-1)^2$

$\dfrac{4x^2 + 7x - 2 = (4x-1)(x+2)}{\text{LCD} = (4x-1)^2(x+2)}$

19. $\dfrac{8}{xy} + \dfrac{2}{xy^2} = \dfrac{8}{xy} \cdot \dfrac{y}{y} + \dfrac{2}{xy^2}$

$= \dfrac{8y+2}{xy^2}$

21. $\dfrac{5}{x^2+5x+6} + \dfrac{2}{x+3}$

$= \dfrac{5}{(x+3)(x+2)} + \dfrac{2}{x+3} \cdot \dfrac{x+2}{x+2}$

$= \dfrac{5+2x+4}{(x+3)(x+2)} = \dfrac{2x+9}{(x+3)(x+2)}$

23. $\dfrac{2}{y-1} + \dfrac{2}{y+1}$

$= \dfrac{2}{y-1} \cdot \dfrac{y+1}{y+1} + \dfrac{2}{y+1} \cdot \dfrac{y-1}{y-1}$

$= \dfrac{2y+2+2y-2}{(y-1)(y+1)} = \dfrac{4y}{(y-1)(y+1)}$

25. $\dfrac{5}{2ab} + \dfrac{1}{2a+b}$

$= \dfrac{5}{2ab} \cdot \dfrac{2a+b}{2a+b} + \dfrac{1}{2a+b} \cdot \dfrac{2ab}{2ab}$

$= \dfrac{10a+5b+2ab}{2ab(2a+b)}$

27. $\dfrac{x-3}{4x} + \dfrac{6}{x^2} = \dfrac{x-3}{4x} \cdot \dfrac{x}{x} + \dfrac{6}{x^2} \cdot \dfrac{4}{4}$

$= \dfrac{x^2 - 3x + 24}{4x^2}$

29. $\dfrac{5}{x^2-25} + \dfrac{1}{x^2+10x+25}$

$= \dfrac{5}{(x+5)(x-5)} \cdot \dfrac{x+5}{x+5} + \dfrac{1}{(x+5)^2} \cdot \dfrac{x-5}{x-5}$

$= \dfrac{5x+25+x-5}{(x+5)^2(x-5)} = \dfrac{6x+20}{(x+5)^2(x-5)}$

31. $\dfrac{3x+5}{x^2+4x+3}+\dfrac{-x+5}{x^2+2x-3}$

$=\dfrac{3x+5}{(x+3)(x+1)}\cdot\dfrac{x-1}{x-1}+\dfrac{-x+5}{(x+3)(x-1)}\cdot\dfrac{x+1}{x+1}$

$=\dfrac{3x^2+2x-5-x^2+4x+5}{(x+3)(x+1)(x-1)}$

$=\dfrac{2x^2+6x}{(x+3)(x+1)(x-1)}=\dfrac{2x(x+3)}{(x+3)(x+1)(x-1)}$

$=\dfrac{2x}{(x+1)(x-1)}$

33. $\dfrac{a+1}{2}-\dfrac{a-1}{3}=\dfrac{a+1}{2}\cdot\dfrac{3}{3}-\dfrac{a-1}{3}\cdot\dfrac{2}{2}$

$=\dfrac{3a+3-2a+2}{2\cdot3}$

$=\dfrac{a+5}{6}$

35. $\dfrac{4x-9}{3x}-\dfrac{3x-8}{4x}$

$=\dfrac{4x-9}{3x}\cdot\dfrac{4}{4}-\dfrac{3x-8}{4x}\cdot\dfrac{3}{3}$

$=\dfrac{16x-36-9x+24}{12x}$

$=\dfrac{7x-12}{12x}$

37. $\dfrac{6}{3x-4}-\dfrac{5}{4x-3}$

$=\dfrac{6}{3x-4}\cdot\dfrac{4x-3}{4x-3}-\dfrac{5}{4x-3}\cdot\dfrac{3x-4}{3x-4}$

$=\dfrac{24x-18-15x+20}{(3x-4)(4x-3)}$

$=\dfrac{9x+2}{(3x-4)(4x-3)}$

39. $\dfrac{1}{x^2-2x}-\dfrac{5}{x^2-4x+4}$

$=\dfrac{1}{x(x-2)}\cdot\dfrac{x-2}{x-2}-\dfrac{5}{(x-2)^2}\cdot\dfrac{x}{x}$

$=\dfrac{x-2-5x}{x(x-2)^2}=\dfrac{-4x-2}{x(x-2)^2}$

41. $\dfrac{2x}{x^2+5x+6}-\dfrac{x+1}{x^2+2x-3}$

$=\dfrac{2x}{(x+2)(x+3)}\cdot\dfrac{x-1}{x-1}-\dfrac{x+1}{(x+3)(x-1)}\cdot\dfrac{x+2}{x+2}$

$=\dfrac{2x^2-2x-(x^2+3x+2)}{(x+3)(x+2)(x-1)}$

$=\dfrac{x^2-5x-2}{(x+3)(x+2)(x-1)}$

43. $\dfrac{3}{ab}+\dfrac{2}{a-b}+\dfrac{a}{b}$

$=\dfrac{3}{ab}\cdot\dfrac{a-b}{a-b}+\dfrac{2}{a-b}\cdot\dfrac{ab}{ab}+\dfrac{a}{b}\cdot\dfrac{a(a-b)}{a(a-b)}$

$=\dfrac{3a-3b+2ab+a^3-a^2b}{ab(a-b)}$

45. $\dfrac{x}{x-5}-\dfrac{2x}{5-x}=\dfrac{x}{x-5}+\dfrac{2x}{x-5}$

$=\dfrac{x+2x}{x-5}=\dfrac{3x}{x-5}$

47. $\dfrac{y-23}{y^2-y-20}+\dfrac{2}{y-5}$

$=\dfrac{y-23}{(y-5)(y+4)}+\dfrac{2}{y-5}\cdot\dfrac{y+4}{y+4}$

$=\dfrac{y-23+2y+8}{(y-5)(y+4)}=\dfrac{3y-15}{(y-5)(y+4)}$

$=\dfrac{3(y-5)}{(y-5)(y+4)}=\dfrac{3}{y+4}$

49. $\dfrac{2}{x^2-9}+\dfrac{x}{x+3}+\dfrac{2x}{x-3}$

$=\dfrac{2}{(x+3)(x-3)}+\dfrac{x}{x+3}\cdot\dfrac{x-3}{x-3}+\dfrac{2x}{x-3}\cdot\dfrac{x+3}{x+3}$

$=\dfrac{2+x^2-3x+2x^2+6x}{(x+3)(x-3)}=\dfrac{3x^2+3x+2}{(x+3)(x-3)}$

51. $\dfrac{2y}{3y^2-8y-3}+\dfrac{1}{6y-2y^2}$

$=\dfrac{2y}{(3y+1)(y-3)}-\dfrac{1}{2y^2-6y}$

$=\dfrac{2y}{(3y+1)(y-3)}\cdot\dfrac{2y}{2y}-\dfrac{1}{2y(y-3)}\cdot\dfrac{3y+1}{3y+1}$

$=\dfrac{4y^2-3y-1}{2y(y-3)(3y+1)}=\dfrac{(4y+1)(y-1)}{2y(3y+1)(y-3)}$

53. $\dfrac{12}{15x-3}+\dfrac{1}{10x-2}=\dfrac{12}{3(5x-1)}+\dfrac{1}{2(5x-1)}$

$=\dfrac{4}{(5x-1)}\cdot\dfrac{2}{2}+\dfrac{1}{2(5x-1)}=\dfrac{9}{2(5x-1)}$

55. $\dfrac{2}{x^2+4x+4}-\dfrac{2x-1}{x^2+5x+6}$

$=\dfrac{2}{(x+2)^2}\cdot\dfrac{x+3}{x+3}-\dfrac{2x-1}{(x+2)(x+3)}\cdot\dfrac{x+2}{x+2}$

$=\dfrac{2x+6-(2x^2+3x-2)}{(x+2)^2(x+3)}$

$=\dfrac{-2x^2-x+8}{(x+2)^2(x+3)}$

57. $\dfrac{42x}{y^2-49}+\dfrac{3x}{y-7}+\dfrac{3x}{y+7}$

$=\dfrac{42x}{(y+7)(y-7)}+\dfrac{3x}{y-7}\cdot\dfrac{y+7}{y+7}+\dfrac{3x}{y+7}\cdot\dfrac{y-7}{y-7}$

$=\dfrac{42x+3xy+21x+3xy-21x}{(y+7)(y-7)}$

$=\dfrac{42x+6xy}{(y+7)(y-7)}=\dfrac{6x(7+y)}{(y+7)(y-7)}$

$=\dfrac{6x}{y-7}$

59. $\dfrac{5x}{x^2+x-6}+\dfrac{2x+1}{x^2+2x-3}-\dfrac{1-2x}{x^2-3x+2}$

$=\dfrac{5x}{(x+3)(x-2)}\cdot\dfrac{x-1}{x-1}+\dfrac{2x+1}{(x+3)(x-1)}\cdot\dfrac{x-2}{x-2}$

$+\dfrac{2x-1}{(x-2)(x-1)}\cdot\dfrac{x+3}{x+3}$

$=\dfrac{5x^2-5x+2x^2-3x-2+2x^2+5x-3}{(x+3)(x-2)(x-1)}$

$=\dfrac{9x^2-3x-5}{(x+3)(x-2)(x-1)}$

Cumulative Review Problems

61. $\quad 5ax=2(ay-3bc)$

$\quad 5ax=2ay-6bc$

$5ax+6bc=2ay$

$\dfrac{5ax+6bc}{2a}=y$

63. $(3x^3y^4)^4=3^4x^{3\cdot4}y^{4\cdot4}$

$=81x^{12}y^{16}$

Putting Your Skills To Work

1. $\qquad x=30,\ y=50$

Avg Speed $=\dfrac{2}{\frac{1}{x}+\frac{1}{y}}=\dfrac{2}{\frac{1}{30}+\frac{1}{50}}$

$=\dfrac{2}{\frac{1}{30}\cdot\frac{5}{5}+\frac{1}{50}\cdot\frac{3}{3}}=\dfrac{2}{\frac{5+3}{150}}$

$=2\cdot\dfrac{150}{8}=37.5$

Average speed $=37.5$ mi / h

3. The student inverted both the numerator and the denominator and treated the inverse of $\dfrac{1}{30}+\dfrac{1}{50}$ as $\dfrac{30+50}{1}$.

101

Exercises 6.4

1. $\dfrac{\frac{3}{x}}{\frac{2}{y}} = \dfrac{3}{x} \cdot \dfrac{y}{2} = \dfrac{3y}{2x}$

3. $\dfrac{\frac{1}{x}+\frac{1}{y}}{\frac{1}{xy}} = \dfrac{\frac{1}{x}+\frac{1}{y}}{\frac{1}{xy}} \cdot \dfrac{(xy)}{(xy)}$

$= \dfrac{\frac{xy}{x}+\frac{xy}{x}}{\frac{xy}{xy}} = \dfrac{y+x}{1} = y+x$

5. $\dfrac{\frac{1}{x}+\frac{1}{y}}{x^2-y^2} = \dfrac{\frac{1}{x}+\frac{1}{y}}{(x+y)(x-y)} \cdot \dfrac{(xy)}{(xy)}$

$= \dfrac{\frac{xy}{x}+\frac{xy}{y}}{(x+y)(x-y)(xy)} = \dfrac{y+x}{(x+y)(x-y)(xy)}$

$= \dfrac{1}{xy(x-y)}$

7. $\dfrac{1-\frac{9}{x^2}}{\frac{3}{x}+1} = \dfrac{1-\frac{9}{x^2}}{\frac{3}{x}+1} \cdot \dfrac{x^2}{x^2}$

$= \dfrac{x^2-\frac{9x^2}{x^2}}{\frac{3x^2}{x}+x^2} = \dfrac{x^2-9}{3x+x^2}$

$= \dfrac{(x+3)(x-3)}{x(3+x)} = \dfrac{x-3}{x}$

9. $\dfrac{\frac{2}{x+1}-2}{3} = \dfrac{\frac{2}{x+1}-2}{3} \cdot \dfrac{(x+1)}{(x+1)}$

$= \dfrac{\frac{2(x+1)}{(x+1)}-2(x+1)}{3(x+1)} = \dfrac{2-2x-2}{3(x+1)} = \dfrac{-2x}{3(x+1)}$

11. $\dfrac{a+\frac{3}{a}}{\frac{a^2+2}{3a}} = \dfrac{a+\frac{3}{a}}{\frac{a^2+2}{3a}} \cdot \dfrac{3a}{3a}$

$= \dfrac{3a^2+\frac{9a}{a}}{\frac{(a^2+2)(3a)}{3a}} = \dfrac{3a^2+9}{a^2+2}$

13. $\dfrac{2+\frac{3y}{x-y}}{2-\frac{3y}{x+2y}} = \dfrac{\frac{2(x-y)}{x-y}+\frac{3y}{x-y}}{\frac{2(x+2y)}{(x+2y)}-\frac{3y}{x+2y}}$

$= \dfrac{\frac{2x-2y+3y}{x-y}}{\frac{2x+4y-3y}{x+2y}} = \dfrac{\frac{2x+y}{x-y}}{\frac{2x+y}{x+2y}}$

$= \dfrac{2x+y}{x-y} \cdot \dfrac{x+2y}{2x+y} = \dfrac{x+2y}{x-y}$

15. $\dfrac{\frac{x}{4}-\frac{1}{2}}{\frac{x}{4}+\frac{1}{x}} = \dfrac{\frac{x}{4}-\frac{1}{2}}{\frac{x}{4}+\frac{1}{x}} \cdot \dfrac{4x}{4x}$

$= \dfrac{\frac{4x^2}{4}-\frac{4x}{2}}{\frac{4x^2}{4}+\frac{4x}{x}} = \dfrac{x^2-2x}{x^2+4} = \dfrac{x(x-2)}{x^2+4}$

17. $\dfrac{\frac{1}{x^2-9}+\frac{2}{x+3}}{\frac{3}{x-3}} = \dfrac{\frac{1}{(x+3)(x-3)}+\frac{2}{x+3}}{\frac{3}{x-3}}$

$= \dfrac{\frac{1}{(x+3)(x-3)}+\frac{2}{x+3}}{\frac{3}{x-3}} \cdot \dfrac{(x+3)(x-3)}{(x+3)(x-3)}$

$= \dfrac{(1)\frac{(x+3)(x-3)}{(x+3)(x-3)}+2\frac{(x+3)(x-3)}{x+3}}{\frac{3(x+3)(x-3)}{x-3}}$

$= \dfrac{1+2(x-3)}{3(x+3)} = \dfrac{1+2x-6}{3(x+3)} = \dfrac{2x-5}{3(x+3)}$

19. $\dfrac{\frac{2}{y-1}+2}{\frac{2}{y+1}-2} = \dfrac{\frac{2}{y-1}+2}{\frac{2}{y+1}-2} \cdot \dfrac{(y-1)(y+1)}{(y-1)(y+1)}$

$= \dfrac{\frac{2(y-1)(y+1)}{y-1}+2(y-1)(y+1)}{\frac{2(y-1)(y+1)}{(y+1)}-2(y-1)(y+1)}$

$= \dfrac{2(y+1)[1+(y-1)]}{2(y-1)[1-(y+1)]} = \dfrac{(y+1)(y)}{(y-1)(1-y-1)}$

$= \dfrac{y(y+1)}{(y-1)(-y)} = \dfrac{y+1}{-y+1}$

21. $\dfrac{\frac{y}{y+1}-\frac{y-1}{y}}{\frac{y+1}{y}+\frac{y}{y-1}} = \dfrac{\frac{y(y)}{(y+1)(y)}-\frac{(y-1)(y+1)}{y(y+1)}}{\frac{(y+1)(y-1)}{y(y-1)}+\frac{y(y)}{(y-1)(y)}}$

$= \dfrac{\frac{y^2-(y^2-1)}{y(y+1)}}{\frac{y^2-1+y^2}{y(y-1)}} = \dfrac{\frac{y^2-y^2+1}{y(y+1)}}{\frac{2y^2-1}{y(y-1)}}$

$= \dfrac{1}{y(y+1)} \cdot \dfrac{y(y-1)}{2y^2-1} = \dfrac{y-1}{(y+1)(2y^2-1)}$

Cumulative Review Problems

23.
$$P = 2(\ell + w)$$
$$P = 2\ell + 2w$$
$$P - 2\ell = 2w$$
$$\frac{P - 2\ell}{2} = w$$

Exercises 6.5

1.
$$\frac{x+1}{2x} = \frac{2}{3}$$
$$6x\left(\frac{x+1}{2x}\right) = 6x\left(\frac{2}{3}\right)$$
$$3(x+1) = 4x$$
$$3x + 3 = 4x$$
$$3 = x$$

Check: $\dfrac{3+1}{2(3)} \overset{?}{=} \dfrac{2}{3}$

$$\frac{4}{6} \overset{?}{=} \frac{2}{3}$$
$$\frac{2}{3} = \frac{2}{3}$$

3.
$$\frac{1}{x} = \frac{1}{4-x}$$
$$x(4-x)\left(\frac{1}{x}\right) = x(4-x)\left(\frac{1}{4-x}\right)$$
$$4 - x = x$$
$$4 = 2x$$
$$2 = x$$

Check: $\dfrac{1}{2} \overset{?}{=} \dfrac{1}{4-2}$

$$\frac{1}{2} = \frac{1}{2}$$

5.
$$\frac{y-3}{3y+2} = \frac{1}{5}$$
$$5(3y+2)\left(\frac{y-3}{3y+2}\right) = 5(3y+2)\left(\frac{1}{5}\right)$$
$$5(y-3) = 3y+2$$
$$5y - 15 = 3y + 2$$
$$2y - 15 = 2$$
$$2y = 17$$
$$y = \frac{17}{2}$$

Check: $\dfrac{\frac{17}{2} - 3}{3\left(\frac{17}{2}\right) + 2} \overset{?}{=} \dfrac{1}{5}$

$$\frac{\frac{11}{2}}{\frac{55}{2}} \overset{?}{=} \frac{1}{5}$$
$$\frac{1}{5} = \frac{1}{5}$$

7.
$$\frac{2}{x} + \frac{x}{x+1} = 1$$
$$x(x+1)\left(\frac{2}{x}\right) + x(x+1)\left(\frac{x}{x+1}\right) = x(x+1)(1)$$
$$(x+1)(2) + x^2 = x^2 + x$$
$$2x + 2 + x^2 = x^2 + x$$
$$2x + 2 = x$$
$$2 = -x$$
$$-2 = x$$

Check: $\dfrac{2}{-2} + \dfrac{-2}{-2+1} \overset{?}{=} 1$

$$-1 + 2 \overset{?}{=} 1$$
$$1 = 1$$

9.
$$\frac{x+1}{x} = 1 + \frac{x-2}{2x}$$
$$2x\left(\frac{x+1}{x}\right) = 2x(1) + 2x\left(\frac{x-2}{2x}\right)$$
$$2(x+1) = 2x + x - 2$$
$$2x + 2 = 3x - 2$$
$$2 = x - 2$$
$$4 = x$$

Check: $\dfrac{4+1}{4} \overset{?}{=} 1 + \dfrac{4-2}{2(4)}$

$$\dfrac{5}{4} \overset{?}{=} \dfrac{10}{8}$$

$$\dfrac{5}{4} = \dfrac{5}{4}$$

11. $\dfrac{x+1}{x} = \dfrac{1}{2} - \dfrac{4}{3x}$

$$6x\left(\dfrac{x+1}{x}\right) = 6x\left(\dfrac{1}{2}\right) - 6x\left(\dfrac{4}{3x}\right)$$

$$6(x+1) = 3x - 8$$

$$6x + 6 = 3x - 8$$

$$3x + 6 = -8$$

$$3x = -14$$

$$x = \dfrac{-14}{3}$$

Check: $\dfrac{-\frac{14}{3}+1}{-\frac{14}{3}} \overset{?}{=} \dfrac{1}{2} - \dfrac{4}{3\left(-\frac{14}{3}\right)}$

$$\dfrac{-\frac{11}{3}}{-\frac{14}{3}} \overset{?}{=} \dfrac{1}{2} + \dfrac{4}{14}$$

$$\dfrac{11}{14} = \dfrac{11}{14}$$

13. $\dfrac{x}{x-2} - 2 = \dfrac{2}{x-2}$

$$(x-2)\left(\dfrac{x}{x-2}\right) - (x-2)(2) = (x-2)\left(\dfrac{2}{x-2}\right)$$

$$x - 2x + 4 = 2$$

$$-x + 4 = 2$$

$$-x = -2$$

$$x = 2$$

But $x = 2$ causes 0 in the denominator of

$\dfrac{x}{x-2}$. $x = 2$ is extraneous and this equation has no solution.

15. $\dfrac{2}{x+1} - \dfrac{1}{x-1} = \dfrac{2x}{x^2-1}$

$$\dfrac{2}{x+1} - \dfrac{1}{x-1} = \dfrac{2x}{(x+1)(x-1)}$$

$$(x+1)(x-1)\left(\dfrac{2}{x+1}\right) - (x+1)(x-1)\left(\dfrac{1}{x-1}\right)$$

$$= (x+1)(x-1)\left(\dfrac{2x}{(x+1)(x-1)}\right)$$

$$(x-1)(2) - (x+1) = 2x$$

$$2x - 2 - x - 1 = 2x$$

$$x - 3 = 2x$$

$$-3 = x$$

Check: $\dfrac{2}{-3+1} - \dfrac{1}{-3-1} \overset{?}{=} \dfrac{2(-3)}{(-3)^2-1}$

$$-1 + \dfrac{1}{4} \overset{?}{=} -\dfrac{6}{8}$$

$$-\dfrac{3}{4} = -\dfrac{3}{4}$$

17. $\dfrac{y+1}{y^2+2y-3} = \dfrac{1}{y+3} - \dfrac{1}{y-1}$

$$\dfrac{y+1}{(y+3)(y-1)} = \dfrac{1}{y+3} - \dfrac{1}{y-1}$$

$$(y+3)(y-1)\left(\dfrac{y+1}{(y+3)(y-1)}\right)$$

$$= (y+3)(y-1)\left(\dfrac{1}{y+3}\right) - (y+3)(y-1)\left(\dfrac{1}{y-1}\right)$$

$$y + 1 = y - 1 - y - 3$$

$$y + 1 = -4$$

$$y = -5$$

Check: $\dfrac{-5+1}{(-5)^2+2(-5)-3} \overset{?}{=} \dfrac{1}{-5+3} - \dfrac{1}{-5-1}$

$$\dfrac{-4}{12} \overset{?}{=} -\dfrac{1}{2} + \dfrac{1}{6}$$

$$-\dfrac{4}{12} = -\dfrac{4}{12}$$

19. $$\frac{79-x}{x}=5+\frac{7}{x}$$

$$x\left(\frac{79-x}{x}\right)=x(5)+x\left(\frac{7}{x}\right)$$

$$79-x=5x+7$$

$$79=6x+7$$

$$72=6x$$

$$12=x$$

Check: $$\frac{79-12}{12}\overset{?}{=}5+\frac{7}{12}$$

$$\frac{67}{12}=\frac{67}{12}$$

21. $$\frac{3}{x+3}+\frac{1}{x-1}=\frac{8}{x^2+2x-3}$$

$$\frac{3}{x+3}+\frac{1}{x-1}=\frac{8}{(x+3)(x-1)}$$

$$(x+3)(x-1)\left(\frac{3}{x+3}\right)+(x+3)(x-1)\left(\frac{1}{x-1}\right)$$

$$=(x+3)(x-1)\left(\frac{8}{(x+3)(x-1)}\right)$$

$$(x-1)(3)+x+3=8$$

$$3x-3+x+3=8$$

$$4x=8$$

$$x=2$$

Check: $$\frac{3}{2+3}+\frac{1}{2-1}\overset{?}{=}\frac{8}{(2)^2+2(2)-3}$$

$$\frac{8}{5}=\frac{8}{5}$$

23. $$\frac{x+11}{x^2-5x+4}+\frac{3}{x-1}=\frac{5}{x-4}$$

$$\frac{x+11}{(x-4)(x-1)}+\frac{3}{x-1}=\frac{5}{x-4}$$

$$(x-4)(x-1)\left(\frac{x+11}{(x-4)(x-1)}\right)+(x-4)(x-1)\left(\frac{3}{x-1}\right)$$

$$=(x-4)(x-1)\left(\frac{5}{x-4}\right)$$

$$x+11+(x-4)(3)=(x-1)(5)$$

$$x+11+3x-12=5x-5$$

$$4x-1=5x-5$$

$$-1=x-5$$

$$4=x \text{ Extraneous}$$

No solution.

25. $$\frac{2x}{x+4}-\frac{8}{x-4}=\frac{2x^2+32}{x^2-16}$$

$$\frac{2x}{x+4}-\frac{8}{x-4}=\frac{2x^2+32}{(x+4)(x-4)}$$

$$(x+4)(x-4)\left(\frac{2x}{x+4}\right)-(x+4)(x-4)\left(\frac{8}{x-4}\right)$$

$$=(x+4)(x-4)\left(\frac{2x^2+32}{(x+4)(x-4)}\right)$$

$$(x-4)(2x)-(x+4)(8)=2x^2+32$$

$$2x^2-8x-8x-32=2x^2+32$$

$$2x^2-16x-32=2x^2+32$$

$$-16x-32=32$$

$$-16x=64$$

$$x=-4 \text{ Extraneous}$$

No solution.

27. $$\frac{23}{5}-\frac{3}{5x-10}=\frac{4}{x-2}$$

$$5(x-2)\left(\frac{23}{5}\right)-5(x-2)\left[\frac{3}{5(x-2)}\right]=5(x-2)\left(\frac{4}{x-2}\right)$$

$$(x-2)(23)-3=20$$

$$23x-46-3=20$$

$$23x=69$$

$$x=3$$

Check: $\dfrac{23}{5} - \dfrac{3}{5(3)-10} \overset{?}{=} \dfrac{4}{3-2}$

$\dfrac{20}{5} \overset{?}{=} 4$

$4 = 4$

29. $\dfrac{2}{x+7} + \dfrac{4x-1}{x^2+5x-14} = \dfrac{1}{x-2}$

$(x+7)(x-2)\left(\dfrac{2}{x+7}\right) + (x+7)(x-2)\left[\dfrac{4x-1}{(x+7)(x-2)}\right]$

$= (x+7)(x-2)\left(\dfrac{1}{x-2}\right)$

$(x-2)(2) + 4x-1 = x+7$

$2x-4+4x-1 = x+7$

$5x = 12$

$x = \dfrac{12}{5}$

Check: $\dfrac{2}{\frac{12}{5}+7} + \dfrac{4\left(\frac{12}{5}\right)-1}{\left(\frac{12}{5}\right)^2 + 5\left(\frac{12}{5}\right) - 14} \overset{?}{=} \dfrac{1}{\frac{12}{5}-2}$

$\dfrac{10}{47} + \dfrac{\frac{43}{5}}{\frac{94}{25}} \overset{?}{=} \dfrac{5}{2}$

$\dfrac{235}{94} \overset{?}{=} \dfrac{5}{2}$

$\dfrac{5}{2} = \dfrac{5}{2}$

31. $\dfrac{10}{x-2} - \dfrac{40}{x^2+x-6} = \dfrac{12}{x+3}$

$\dfrac{10}{x-2} - \dfrac{40}{(x+3)(x-2)} = \dfrac{12}{x+3}$

$(x-2)(x+3)\left(\dfrac{10}{x-2}\right) - (x-2)(x+3)\left(\dfrac{40}{(x+3)(x-2)}\right)$

$= (x-2)(x+3)\left(\dfrac{12}{x+3}\right)$

$(x+3)(10) - 40 = (x-2)(12)$

$10x+30-40 = 12x-24$

$10x-10 = 12x-24$

$-10 = 2x-24$

$14 = 2x$

$7 = x$

Check: $\dfrac{10}{7-2} - \dfrac{40}{(7)^2+7-6} \overset{?}{=} \dfrac{12}{7+3}$

$2 - \dfrac{4}{5} \overset{?}{=} \dfrac{12}{10}$

$\dfrac{6}{5} = \dfrac{6}{5}$

33. $\dfrac{3x}{x-2} - \dfrac{4x}{x-4} = \dfrac{3}{x^2-6x+8}$

$x-2 \neq 0 \Rightarrow x \neq 2$

$x-4 \neq 0 \Rightarrow x \neq 4$

$x^2-6x+8 \neq 0$

$(x-2)(x-4) \neq 0$

$x-2 \neq 0 \qquad x-4 \neq 0$

$x \neq 2 \qquad\quad x \neq 4$

Extraneous solutions $x = 2$, $x = 4$.

Cumulative Review Problems

35. $6x^2 - x - 12 = (3x+4)(2x-3)$

37. $x = \text{width}$

$3x-1 = \text{length}$

$P = 2L + 2W$

$54 = 2(3x-1) + 2x$

$54 = 6x-2+2x$

$56 = 8x$

$7 = x$

$20 = 3x-1$

Width = 7 meters and Length = 20 meters

Exercises 6.6

1. $\dfrac{7}{5} = \dfrac{21}{x}$

$7x = 105$

$x = 15$

3. $\dfrac{x}{17} = \dfrac{12}{5}$

$5x = 204$

$x = \dfrac{204}{5}$ or $40\dfrac{4}{5}$

5. $\dfrac{5}{3} = \dfrac{x}{8}$

$40 = 3x$

$\dfrac{40}{3} = x$ or $x = 13\dfrac{1}{3}$

7. $\dfrac{5}{x} = \dfrac{150}{75}$

$375 = 150x$

$\dfrac{375}{150} = x$

$\dfrac{5}{2} = x$ or $x = 2\dfrac{1}{2}$

9. x = cups from 3 pounds

$\dfrac{x}{3} = \dfrac{190}{5}$

$5x = 570$

$x = 114$

It will yield 114 cups.

11. x = gallons of gas for 11 pints of oil

$\dfrac{x}{11} = \dfrac{7}{4}$

$4x = 77$

$x = \dfrac{77}{4}$ or $19\dfrac{1}{4}$

He should use $19\dfrac{1}{4}$ gallons.

13. x = weight of 60 yards

$\dfrac{x}{60} = \dfrac{210}{25}$

$25x = 12,600$

$x = 504$

It weighs 504 pounds.

15. x = speed limit in miles per hour

$\dfrac{x}{90} = \dfrac{62}{100}$

$100x = 5580$

$x = 55.8 \approx 56$

The limit is 56 miles per hour.

17. x = gallons of gas to go from Boston to Chicago

$\dfrac{x}{970} = \dfrac{8}{170}$

$170x = 7760$

$x = 45.6 \approx 46$

They will use 46 gallons.

19. x = miles from base of mountain

$\dfrac{x}{\frac{3}{4}} = \dfrac{136}{3\frac{1}{2}}$

$\dfrac{7}{2}x = \dfrac{3}{4}(136)$

$\dfrac{7}{2}x = 102$

$x = 29.1$

He is 29 miles away.

21. $x = 20, \quad y = 29, \quad m = 13$

$\dfrac{n}{m} = \dfrac{y}{x}$

$\dfrac{n}{13} = \dfrac{29}{20}$

$20n = 377$

$n = \dfrac{377}{20} = 18\dfrac{17}{20}$ inches

23. $n = 40, \quad m = 35, \quad x = 100$

$\dfrac{y}{x} = \dfrac{n}{m}$

$\dfrac{y}{100} = \dfrac{40}{35}$

$35y = 4000$

$y = \dfrac{800}{7} = 114\dfrac{2}{7}$ meters

25. x = height of building

$\dfrac{x}{47} = \dfrac{6}{5}$

$5x = 282$

$x = 56.4 \approx 56$

The building is 56 feet tall.

27. x = height of kite on long line

$$\frac{x}{120} = \frac{5}{7}$$

$$7x = 600$$

$$x = 85.7 \approx 86$$

The kite is 86 meters high.

29.

	D	R	T = D/R
Speedboat A	240	s + 10	240/s + 10
Speedboat B	210	s	210/s

$$\frac{240}{s+10} = \frac{210}{s}$$

$$240s = 210s + 2100$$

$$30s = 2100$$

$$s = 70$$

$$s + 10 = 80$$

Speedboat A went 80 km / h.

Speedboat B went 70 km / h.

31.

	D	R	T = D/R
Helicopter A	1050	s	1050/s
Helicopter B	1250	s + 40	1250/s+40

$$\frac{1250}{s+40} = \frac{1050}{s}$$

$$1250s = 1050s + 42{,}000$$

$$200s = 42{,}000$$

$$s = 210$$

$$s + 40 = 250$$

Helicopter A flies at 210 km / h.

Helicopter B flies at 250 km / h.

33. x = time to mow together

$$\frac{1}{4} + \frac{1}{5} = \frac{1}{x}$$

$$5x + 4x = 20$$

$$9x = 20$$

$$x = \frac{20}{9} \text{ or } x = 60\left(\frac{20}{9}\right) = 133.3$$

It will take 133 minutes or 2 hours 13 minutes.

35. x = time for pumps together

$$\frac{1}{15} + \frac{1}{20} = \frac{1}{x}$$

$$4x + 3x = 60$$

$$7x = 60$$

$$x = \frac{60}{7} \text{ or } x = 60\left(\frac{60}{7}\right) = 514.3$$

It will take 514 minutes or 8 hours 34 minutes.

Cumulative Review Problems

37. $0.0000006316 = 6.316 \times 10^{-7}$

39. $\dfrac{x^{-3}y^{-2}}{z^4 w^{-8}} = \dfrac{w^8}{x^3 y^2 z^4}$

Putting Your Skills To Work

1. $x = 10$ and $y = 6$

$$\frac{1}{R} = \frac{1}{x} + \frac{1}{y}$$

$$\frac{1}{R} = \frac{1}{10} + \frac{1}{6}$$

$$30 = 3R + 5R$$

$$30 = 8R$$

$$\frac{15}{4} = R$$

Resistance is 3.75 ohms.

3. $x = 8, \ y = 12, \ w = 6, \ z = 3$

$$\frac{1}{R} = \frac{1}{x} + \frac{1}{y} + \frac{1}{w} + \frac{1}{z}$$

$$\frac{1}{R} = \frac{1}{8} + \frac{1}{12} + \frac{1}{6} + \frac{1}{3}$$

$$24 = 3R + 2R + 4R + 8R$$

$$24 = 17R$$

$$\frac{24}{17} = 1.4$$

Resistance is 1.4 ohms.

1. $\dfrac{4x-4y}{5y-5x} = \dfrac{4(x-y)}{-5(x-y)} = -\dfrac{4}{5}$

3. $\dfrac{2x^2+5x-3}{2x^2-9x+4} = \dfrac{(2x-1)(x+3)}{(2x-1)(x-4)} = \dfrac{x+3}{x-4}$

5. $\dfrac{x^2-9}{x^2-10x+21} = \dfrac{(x+3)(x-3)}{(x-7)(x-3)} = \dfrac{x+3}{x-7}$

7. $\dfrac{4x^2+4x+3}{4x^2-2x} = \dfrac{(2x+3)(2x-1)}{2x(2x-1)} = \dfrac{2x+3}{2x}$

9. $\dfrac{2x^2-2xy-24y^2}{2x^2+5xy-3y^2} = \dfrac{(2x-8y)(x+3y)}{(2x-y)(x+3y)} = \dfrac{2x-8y}{2x-y}$

11. $\dfrac{5x^3-10x^2}{25x^4+5x^3-30x^2} = \dfrac{5x^2(x-2)}{5x^2(5x^2+x-6)} = \dfrac{x-2}{5x^2+x-6}$

13. $\dfrac{3x^2-13x-10}{3x^2+2x} \cdot \dfrac{x^2-25x}{x^2-25}$

$= \dfrac{(3x+2)(x-5)}{x(3x+2)} \cdot \dfrac{x(x-25)}{(x+5)(x-5)}$

$= \dfrac{x-25}{x+5}$

15. $\dfrac{2x^2-18}{3y^2+3y} + \dfrac{y^2+6y+9}{y^2+4y+3}$

$= \dfrac{2x^2-18}{3y^2+3y} \cdot \dfrac{y^2+4y+3}{y^2+6y+9}$

$= \dfrac{2(x+3)(x-3)}{3y(y+1)} \cdot \dfrac{(y+3)(y+1)}{(y+3)(y+3)}$

$= \dfrac{2(x+3)(x-3)}{3y(y+3)}$

17. $\dfrac{x^3-36x}{12x^2+2x} \cdot \dfrac{36x+6}{x^3-6x^2} \cdot \dfrac{x^2+2x}{x^2+11x+30}$

$= \dfrac{x(x+6)(x-6)}{2x(6x+1)} \cdot \dfrac{6(6x+1)}{x^2(x-6)} \cdot \dfrac{x(x+2)}{(x+5)(x+6)}$

$= \dfrac{3(x+2)}{x(x+5)}$

19. $\dfrac{6y^2+13y-5}{9y^2+3y} + \dfrac{4y^2+20y+25}{12y^2}$

$= \dfrac{6y^2+13y-5}{9y^2+3y} \cdot \dfrac{12y^2}{4y^2+20y+25}$

$= \dfrac{(3y-1)(2y+5)}{3y(3y+1)} \cdot \dfrac{12y^2}{(2y+5)(2y+5)}$

$= \dfrac{4y(3y-1)}{(3y+1)(2y+5)}$

21. $\dfrac{11}{x-2} \cdot \dfrac{2x^2-8}{44} = \dfrac{11}{x-2} \cdot \dfrac{2(x+2)(x-2)}{44} = \dfrac{x+2}{2}$

23. $\dfrac{5x^7}{16x^2-9} \cdot \dfrac{8x+6}{10x^8} = \dfrac{5x^7}{(4x+3)(4x-3)} \cdot \dfrac{2(4x+3)}{10x^8}$

$= \dfrac{1}{x(4x-3)}$

25. $\dfrac{7}{x+1} + \dfrac{4}{2x} = \dfrac{7}{x+1} \cdot \dfrac{2x}{2x} + \dfrac{4}{2x} \cdot \dfrac{x+1}{x+1}$

$= \dfrac{14x+4x+4}{2x(x+1)}$

$= \dfrac{18x+4}{2x(x+1)}$

$= \dfrac{9x+2}{x(x+1)}$

27. $\dfrac{2}{x^2-9} + \dfrac{x}{x+3} = \dfrac{2}{(x+3)(x-3)} + \dfrac{x}{x+3} \cdot \dfrac{x-3}{x-3}$

$= \dfrac{2+x^2-3x}{(x+3)(x-3)}$

$= \dfrac{(x-2)(x-1)}{(x+3)(x-3)}$

29. $\dfrac{x}{y} + \dfrac{3}{2y} + \dfrac{1}{y+2}$

$= \dfrac{x}{y} \cdot \dfrac{2(y+2)}{2(y+2)} + \dfrac{3}{2y} \cdot \dfrac{y+2}{y+2} + \dfrac{1}{y+2} \cdot \dfrac{2y}{2y}$

$= \dfrac{2x(y+2)+3y+6+2y}{2y(y+2)}$

$= \dfrac{2xy+4x+5y+6}{2y(y+2)}$

31.
$$\frac{3x+1}{3x} - \frac{1}{x} = \frac{3x+1}{3x} - \frac{1}{x} \cdot \frac{3}{3}$$
$$= \frac{3x+1-3}{3x}$$
$$= \frac{3x-2}{3x}$$

33.
$$\frac{1}{x^2+7x+10} - \frac{x}{x+5}$$
$$= \frac{1}{(x+2)(x+5)} - \frac{x}{x+5} \cdot \frac{x+2}{x+2}$$
$$= \frac{1-x^2-2x}{(x+2)(x+5)}$$

35.
$$\frac{2x+1}{x^2+5x+6} + \frac{4x}{x^2-9}$$
$$= \frac{2x+1}{(x+3)(x+2)} + \frac{4x}{(x+3)(x-3)}$$
$$= \frac{2x^2-5x-3+4x^2+8x}{(x+3)(x-3)(x+2)}$$
$$= \frac{6x^2+3x-3}{(x+3)(x-3)(x+2)} \text{ or}$$
$$= \frac{3(2x-1)(x+1)}{(x+3)(x-3)(x+2)}$$

37.
$$\frac{\frac{3}{2y} - \frac{1}{y}}{\frac{4}{y} + \frac{3}{2y}} = \frac{\frac{3}{2y} - \frac{1}{y}}{\frac{4}{y} + \frac{3}{2y}} \cdot \frac{2y}{2y}$$
$$= \frac{3-2}{8+3} = \frac{1}{11}$$

39.
$$\frac{w - \frac{4}{w}}{1 + \frac{2}{w}} = \frac{w - \frac{4}{w}}{1 + \frac{2}{w}} \cdot \frac{w}{w}$$
$$= \frac{w^2-4}{w+2} = \frac{(w+2)(w-2)}{w+2}$$
$$= w-2$$

41.
$$\frac{1 + \frac{1}{y^2-1}}{\frac{1}{y+1} - \frac{1}{y-1}} = \frac{1 + \frac{1}{(y+1)(y-1)}}{\frac{1}{y+1} - \frac{1}{y-1}} \cdot \frac{(y+1)(y-1)}{(y+1)(y-1)}$$
$$= \frac{(y+1)(y-1)+1}{y-1-(y+1)}$$
$$= \frac{y^2}{-2} = -\frac{y^2}{2}$$

43.
$$\frac{\frac{1}{a+b} - \frac{1}{a}}{b} = \frac{\frac{a-(a+b)}{a(a+b)}}{b}$$
$$= -\frac{1}{a(a+b)}$$

45.
$$\left(\frac{1}{x+2y} - \frac{1}{x-y}\right) \div \frac{2x-4y}{x^2-3xy+2y^2}$$
$$= \left(\frac{1}{x+2y} - \frac{1}{x-y}\right) \cdot \frac{x^2-3xy+2y^2}{2x-4y}$$
$$= \frac{x-y-(x+2y)}{(x+2y)(x-y)} \cdot \frac{(x-y)(x-2y)}{2(x-2y)}$$
$$= \frac{-3y}{2(x+2y)}$$

47.
$$\frac{8}{a-3} = \frac{12}{a+3}$$
$$(a-3)(a+3)\frac{8}{a-3} = (a-3)(a+3)\frac{12}{a+3}$$
$$8a+24 = 12a-36$$
$$60 = 4a$$
$$15 = a$$

49.
$$\frac{2x-1}{x} - \frac{1}{2} = -2$$
$$2x\left(\frac{2x-1}{x}\right) - 2x\left(\frac{1}{2}\right) = -2(2x)$$
$$4x-2-x = -4x$$
$$7x = 2$$
$$x = \frac{2}{7}$$

51.
$$\frac{5}{2} - \frac{2y+7}{y+6} = 3$$
$$2(y+6)\left(\frac{5}{2}\right) - 2(y+6)\left(\frac{2y+7}{y+6}\right) = 2(y+6)(3)$$
$$5y+30-4y-14 = 6y+36$$
$$-20 = 5y$$
$$-4 = y$$

110

53.
$$\frac{1}{3x} + 2 = \frac{5}{6x} - \frac{1}{2}$$
$$6x\left(\frac{1}{3x}\right) + 6x(2) = 6x\left(\frac{5}{6x}\right) - 6x\left(\frac{1}{2}\right)$$
$$2 + 12x = 5 - 3x$$
$$15x = 3$$
$$x = \frac{1}{5}$$

55.
$$\frac{x-8}{x-2} = \frac{2x}{x+2} - 2$$
$$(x+2)(x-2)\left(\frac{x-8}{x-2}\right)$$
$$= (x+2)(x-2)\left(\frac{2x}{x+2}\right) - (x+2)(x-2)(2)$$
$$x^2 - 6x - 16 = 2x^2 - 4x - 2x^2 + 8$$
$$x^2 - 2x - 24 = 0$$
$$(x-6)(x+4) = 0$$
$$x - 6 = 0 \qquad x + 4 = 0$$
$$x = 6 \qquad x = -4$$

57.
$$\frac{3y-1}{3y} - \frac{6}{5y} = \frac{1}{y} - \frac{4}{15}$$
$$15y\left(\frac{3y-1}{3y}\right) - 15y\left(\frac{6}{5y}\right) = 15y\left(\frac{1}{y}\right) - 15y\left(\frac{4}{15}\right)$$
$$15y - 5 - 18 = 15 - 4y$$
$$19y = 38$$
$$y = 2$$

59.
$$\frac{4}{x^2-1} = \frac{2}{x-1} + \frac{2}{x+1}$$
$$(x+1)(x-1)\left[\frac{4}{(x+1)(x-1)}\right]$$
$$= (x+1)(x-1)\left(\frac{2}{x-1}\right) + (x+1)(x-1)\left(\frac{2}{x+1}\right)$$
$$4 = 2x + 2 + 2x - 2$$
$$4 = 4x$$
$$1 = x$$

$x = 1$ is not an allowed solution.
There is no solution.

61.
$$\frac{9y-3}{y^2+2y} - \frac{5}{y+2} = \frac{3}{y}$$
$$y(y+2)\left[\frac{9y-3}{y(y+2)}\right] - y(y+2)\left(\frac{5}{y+2}\right) = y(y+2)\left(\frac{3}{y}\right)$$
$$9y - 3 - 5y = 3y + 6$$
$$y = 9$$

63.
$$\frac{8}{5} = \frac{2}{x}$$
$$8x = 10$$
$$x = \frac{5}{4} = 1.25 = 1.3$$

65.
$$\frac{33}{10} = \frac{x}{8}$$
$$264 = 10x$$
$$26.4 = x$$

67. $x = $ time to register 550 students
$$\frac{x}{550} = \frac{9}{450}$$
$$450x = 4950$$
$$x = 11$$

It takes 11 hours

69. $x = $ gallons to cover 400 sq. feet
$$\frac{x}{400} = \frac{5}{240}$$
$$240x = 2000$$
$$x = \frac{2000}{240} = 8\frac{1}{3}$$

8.3 gallons of paint are needed

71. $x = $ gallons to get to Denver
$$\frac{x}{1300} = \frac{7}{200}$$
$$200x = 9100$$
$$x = \frac{91}{2} = 45\frac{1}{2}$$

He will use 46 gallons of gas

73.

	D	R	T = D/R
Train	180	S + 20	180/(S+20)
Car	120	S	120/S

$$\frac{180}{s+20} = \frac{120}{s}$$
$$180s = 120s + 2400$$
$$60s = 2400$$
$$s = 40$$
$$s + 20 = 60$$

Car's speed is 40 mph

Trains speed is 60 mph

75. x = height of building

$$\frac{x}{450} = \frac{8}{3}$$
$$3x = 3600$$
$$x = 1200$$

Office building is 1200 ft tall

Test - Chapter 6

1. $\dfrac{2ac + 2ad}{3a^2c + 3a^2d} = \dfrac{2a(c+d)}{3a^2(c+d)} = \dfrac{2}{3a}$

3. $\dfrac{x^2 + 2x}{2x - 1} \cdot \dfrac{10x^2 - 5x}{12x^3 + 24x^2} = \dfrac{x(x+2)}{2x - 1} \cdot \dfrac{5x(2x-1)}{12x^2(x+2)}$
$$= \frac{5}{12}$$

5. $\dfrac{2a^2 - 3a - 2}{4a^2 + a - 14} \div \dfrac{2a^2 + 5a + 2}{16a^2 - 49}$
$$= \frac{2a^2 - 3a - 2}{4a^2 + a - 14} \cdot \frac{16a^2 - 49}{2a^2 + 5a + 2}$$
$$= \frac{(2a+1)(a-2)}{(4a-7)(a+2)} \cdot \frac{(4a+7)(4a-7)}{(2a+1)(a+2)}$$
$$= \frac{(a-2)(4a+7)}{(a+2)^2}$$

7. $\dfrac{x-y}{xy} - \dfrac{a-y}{ay} = \dfrac{x-y}{xy} \cdot \dfrac{a}{a} - \dfrac{a-y}{ay} \cdot \dfrac{x}{x}$
$$= \frac{ax - ay - ax + xy}{axy}$$
$$= \frac{xy - ay}{axy}$$
$$= \frac{y(x-a)}{axy}$$
$$= \frac{x-a}{ax}$$

9. $\dfrac{\frac{x}{3y} - \frac{1}{2}}{\frac{4}{3y} - \frac{2}{x}} = \dfrac{\frac{x}{3y} - \frac{1}{2}}{\frac{4}{3y} - \frac{2}{x}} \cdot \dfrac{6xy}{6xy}$
$$= \frac{2x^2 - 3xy}{8x - 12y}$$
$$= \frac{x(2x - 3y)}{4(2x - 3y)}$$
$$= \frac{x}{4}$$

11. $\dfrac{2x^2 + 3xy - 9y^2}{4x^2 + 13xy + 3y^2} = \dfrac{(2x-3y)(x+3y)}{(4x+y)(x+3y)} = \dfrac{2x-3y}{4x+y}$

13.
$$\frac{15}{x} + \frac{9x - 7}{x + 2} = 9$$
$$x(x+2)\left(\frac{15}{x}\right) + x(x+2)\left(\frac{9x-7}{x+2}\right) = x(x+2)(9)$$
$$15x + 30 + 9x^2 - 7x = 9x^2 + 18x$$
$$30 = 10x$$
$$3 = x$$

15.
$$3 - \frac{7}{x+3} = \frac{x-4}{x+3}$$
$$(x+3)(3) - (x+3)\left(\frac{7}{x+3}\right) = (x+3)\left(\frac{x-4}{x+3}\right)$$
$$3x + 9 - 7 = x - 4$$
$$2x = -6$$
$$x = -3$$

$x = -3$ is not allowed.

There is no solution.

17. $\dfrac{9}{x} = \dfrac{13}{5}$

$45 = 13x$

$\dfrac{45}{13} = x$

19. $x =$ Cost of wood for 92 days

$\dfrac{x}{92} = \dfrac{100}{25}$

$25x = 9200$

$x = 368$

It will cost \$368.

Cumulative Test - Chapters 0 - 6

1. 0.25% of $2.57 = (0.0025)(2.57)$

$\qquad = 0.006$ centimeters

3. $\dfrac{1}{4}\%$ of $19{,}990 = (0.0025)(19{,}990)$

$\qquad = \$49.98$

5. $A = \pi r^2 h$

$\dfrac{A}{\pi r^2} = \dfrac{\pi r^2 h}{\pi r^2}$

$\dfrac{A}{\pi r^2} = h$

7. $\qquad \dfrac{1}{4}x + \dfrac{3}{4} < \dfrac{2}{3}x - \dfrac{4}{3}$

$12\left(\dfrac{1}{4}x\right) + 12\left(\dfrac{3}{4}\right) < 12\left(\dfrac{2}{3}x\right) - 12\left(\dfrac{4}{3}\right)$

$3x + 9 < 8x - 16$

$9 < 5x - 16$

$25 < 5x$

$5 < x$

$x > 5$

9. $8a^3 - 38a^2 b - 10ab^2$

$= 2a\left(4a^2 - 19ab - 5b^2\right)$

$= 2a(4a + b)(a - 5b)$

11. $\dfrac{4x^2 - 25}{2x^2 + 9x - 35} = \dfrac{(2x+5)(2x-5)}{(2x-5)(x+7)}$

$\qquad = \dfrac{2x+5}{x+7}$

13. $\dfrac{5}{2x+4} + \dfrac{3}{x-3} = \dfrac{5}{2(x+2)} \cdot \dfrac{x-3}{x-3} + \dfrac{3}{x-3} \cdot \dfrac{2(x+2)}{2(x+2)}$

$\qquad = \dfrac{5x - 15 + 6x + 12}{2(x+2)(x-3)}$

$\qquad = \dfrac{11x - 3}{2(x+2)(x-3)}$

15. $\qquad \dfrac{3x-2}{3x+2} = 2$

$(3x+2)\left(\dfrac{3x-2}{3x+2}\right) = (3x+2)(2)$

$3x - 2 = 6x + 4$

$-6 = 3x$

$-2 = x$

17. $\dfrac{\frac{1}{x-3} + \frac{5}{x^2-9}}{\frac{6x}{x+3}} = \dfrac{\frac{1}{x-3} + \frac{5}{(x+3)(x-3)}}{\frac{6x}{x+3}} \cdot \dfrac{(x+3)(x-3)}{(x+3)(x-3)}$

$\qquad = \dfrac{x+3+5}{6x(x-3)}$

$\qquad = \dfrac{x+8}{6x(x-3)}$

19. $\dfrac{5}{7} = \dfrac{14}{x}$

$5x = 98$

$x = \dfrac{98}{5}$

21. $\qquad x =$ phone calls to make 110 sales

$\dfrac{x}{110} = \dfrac{22}{5}$

$5x = 2420$

$x = 484$

He expects to make 484 calls.

Practice Quiz: Sections 6.1 - 6.3

1. Simplify $\dfrac{2x^2 + 5x - 3}{x^2 - 2x - 15}$

2. Simplify $\dfrac{9 - x^2}{x^2 - 6x + 9}$

3. Multiply $\dfrac{2x^2 - x - 3}{2x^2 - 2x} \cdot \dfrac{12x^2 - 20x}{6x^2 - 19x + 15}$

4. Add $\dfrac{3}{x - 7} + \dfrac{2x}{x^2 - 49}$

5. Subtract $\dfrac{4}{x} - \dfrac{5}{2x^2 + 3x}$

Practice Quiz: Sections 6.4 - 6.6

1. Simplify $\dfrac{1 - \frac{4}{x^2}}{\frac{2}{x} + 1}$

2. Solve $\dfrac{2}{x - 1} + \dfrac{5}{2x - 2} = \dfrac{3}{4}$

3. Solve $\dfrac{2x}{x - 3} + \dfrac{2}{x - 5} = \dfrac{3x}{x^2 - 8x + 15}$

4. Solve $\dfrac{3}{x} = \dfrac{2}{7}$

5. Solve $\dfrac{150}{s + 2} = \dfrac{125}{s}$

Answers to Practice Quiz

Sections 6.1 - 6.3

1. $\dfrac{2x - 1}{x - 5}$

2. $-\dfrac{x + 3}{x - 3}$

3. $\dfrac{2(x + 1)}{x - 1}$

4. $\dfrac{5x + 21}{(x + 7)(x - 7)}$

5. $\dfrac{8x + 7}{x(2x + 3)}$

Answers to Practice Quiz

Sections 6.4 - 6.6

1. $\dfrac{x - 2}{x}$

2. 7

3. $-\dfrac{1}{2}$, 6

4. $\dfrac{21}{2}$

5. 10

114

1. $(-3, 1)(4, -4)(-2, -6)$

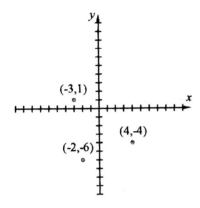

3. $y = -3x + 7$

  (a) $(-1,)$, $y = -3(-1) + 7 = 10$
  $(-1, 10)$

  (b) $(, 1)$, $1 = -3x + 7 \Rightarrow x = 2$
  $(2, 1)$

5. $y = 4$

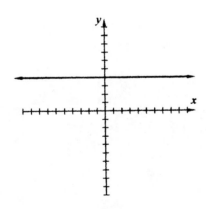

7. $4x - 3y - 7 = 0$
$$-3y = -4x - 7$$
$$y = \frac{4}{3}x + \frac{7}{3}$$
$$y = mx + b$$

$$m = \frac{4}{3}$$

9. $y = 2x + 1$

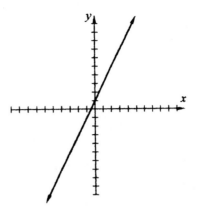

11. $m = -\frac{1}{4}$

  (a) parallel: $m_2 = m_1 = -\frac{1}{4}$

  (b) perpendicular: $m_2 = -\frac{1}{m_1} = 4$

13. $(-1, 6)$ and $(2, 3)$
$$m = \frac{y_2 - y_1}{x_2 - x_1} = \frac{3 - 6}{2 - (-1)} = -1$$
$$y - y_1 = m(x - x_1)$$
$$y - 6 = -1[x - (-1)]$$
$$y = -x + 5$$

15. $4x + 2y < -12$
$$2x + y < -6$$

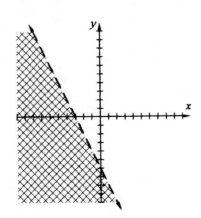

17. Not a function

19. $g(x) = 2x^2$

(a) $g(-6) = 2(-6)^2 = 72$

(b) $g\left(\dfrac{1}{2}\right) = 2\left(\dfrac{1}{2}\right)^2 = \dfrac{1}{2}$

Exercises 7.1

1.

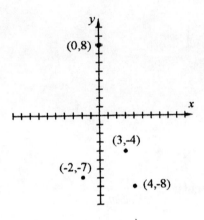

3.

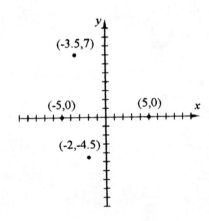

5. R: $(-3, -5)$

S: $\left(-4\dfrac{1}{2}, 0\right)$

X: $(3, -5)$

Y: $\left(2\dfrac{1}{2}, 6\right)$

7. The x – coordinate of the origin is 0.

9. $(5, 1)$ is an ordered pair because the order is important. The graphs of $(5, 1)$ and $(1, 5)$ are different.

11. $(-5, -3), (-4, -4), (-4, -5), (-3, -3)$
$(-2, -1), (-2, -2), (-1, -3), (0, -4)$
$(0, -5), (1, -3)$

13. $y = 3x + 8$

(a) $(0, \ )$
$y = 3(0) + 8 = 8$
$(0, 8)$

(b) $(4, \ )$
$y = 3(4) + 8 = 20$
$(4, 20)$

15. $y = -2x + 3$

(a) $(-6, \ )$
$y = -2(-6) + 3 = 15$
$(-6, 15)$

(b) $(3, \ )$
$y = -2(3) + 3 = -3$
$(3, -3)$

17.　$2x - 8y = 16$

(a) $(8, \ )$

$2(8) - 8y = 16$

$-8y = 0$

$y = 0$

$(8, 0)$

(b) $(\ , -5)$

$2x - 8(-5) = 16$

$2x = -24$

$x = -12$

$(-12, -5)$

19.　$5x + 2y = -20$

(a) $(2, \ )$

$5(2) + 2y = -20$

$2y = -30$

$y = -15$

$(2, -15)$

(b) $(\ , -5)$

$5x + 2(-5) = -20$

$5x = -10$

$x = -2$

$(-2, -5)$

Cumulative Review Problems

21. $\dfrac{8x}{x+y} + \dfrac{8y}{x+y} = \dfrac{8x + 8y}{x+y} = \dfrac{8(x+y)}{x+y} = 8$

23. $r = 20$

$A = \pi r^2 = 3.14(20)^2 = 1256$

The area is 1256 yd^2.

Exercises 7.2

1. Replacing x by -2 and y by 5 in $2x + 5y = 0$ does not result in a true statement so $(-2, 5)$ is not a solution.

3. The x – intercept of a line is the point where the line crosses the _x - axis_.

5. $y = -2x - 3$

$(-2, 1), (-1, -1), (0, -3)$

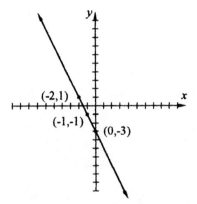

7. $y = x - 4$

$(0, -4), (2, -2), (4, 0)$

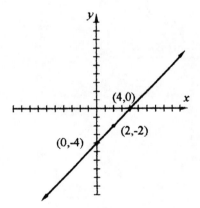

9. $y = 3x - 1$

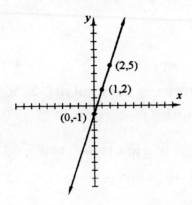

11. $y = 2x + 1$

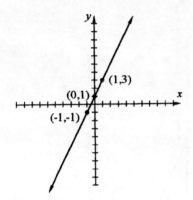

13. $2y - 4x = 0$

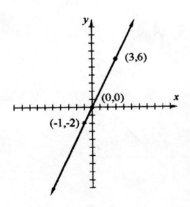

15. $y = 4 - 2x$

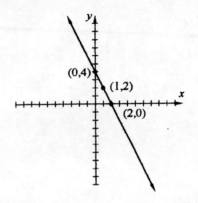

17. $3x + 2y = 6$

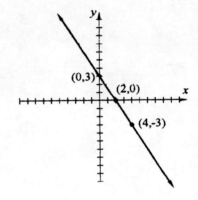

19. $x - 2 = 4y$

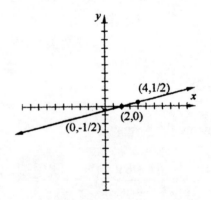

21. $y - 2 = 3y$

118

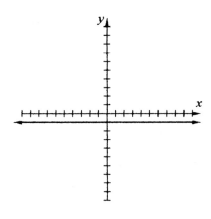

Cumulative Review Problems

23 .

$$\frac{x}{x+4}+\frac{x-1}{x-4}=2$$

$$(x-4)(x)+(x+4)(x-1)=2(x+4)(x-4)$$

$$x^2-4x+x^2+3x-4=2x^2-32$$

$$-x=-28$$

$$x=28$$

25. $x=$ length

$$\frac{x}{2}+1=\text{width}$$

$$p=2L+2w$$

$$53=2x+2\left(\frac{x}{2}+1\right)$$

$$53=2x+x+2$$

$$51=3x$$

$$17=x$$

$$\frac{17}{2}+1=9\frac{1}{2}$$

$$\text{width}=9\frac{1}{2}\text{ meters}$$

$$\text{Length}=17\text{ meters}$$

Use $m=\dfrac{y_2-y_1}{x_2-x_1}$ in Exercises 1-13.

1. $(7,3)$ and $(6,9)$

$$m=\frac{9-3}{6-7}=\frac{6}{-1}=-6$$

3. $(-2,1)$ and $(3,4)$

$$m=\frac{4-1}{3-(-2)}=\frac{3}{5}$$

5. $(-7,-6)$ and $(-3,2)$

$$m=\frac{2-(-6)}{-3-(-7)}=\frac{8}{4}=2$$

7. $(3,-4)$ and $(-6,-4)$

$$m=\frac{-4-(-4)}{-6-3}=\frac{0}{-9}=0$$

9. $(-3,0)$ and $(0,-4)$

$$m=\frac{-4-0}{0-(-3)}=\frac{-4}{3}=-\frac{4}{3}$$

11. $\left(\dfrac{1}{3},-2\right)$ and $\left(\dfrac{2}{3},-5\right)$

$$m=\frac{-5-(-2)}{\frac{2}{3}-\frac{1}{3}}=\frac{-3}{\frac{1}{3}}=-9$$

13. You cannot find the slope of the line passing through $(5,-12)$ and $(5,-6)$ because division by zero is impossible.

Use $y=mx+b$ in Exercises 15-38.

15. $y=8x+9$; $m=8$, $b=9$

17. $y=3x-7$, $m=3$, $b=-7$

19. $y=\dfrac{5}{6}x-\dfrac{2}{9}$; $m=\dfrac{5}{6}$, $b=-\dfrac{2}{9}$

21. $y=-6x$; $m=-6$, $b=0$

23. $4x + y = -\dfrac{1}{3}$

   $y = -4x - \dfrac{1}{3}$; $m = -4$, $b = -\dfrac{1}{3}$

25. $3x - y = -2$

   $y = 3x + 2$; $m = 3$, $b = 2$

27. $5x + 2y = 3$

   $y = -\dfrac{5}{2}x + \dfrac{3}{2}$; $m = -\dfrac{5}{2}$, $b = \dfrac{3}{2}$

29. $4x - 3y = 7$

   $y = \dfrac{4}{3}x - \dfrac{7}{3}$; $m = \dfrac{4}{3}$, $b = -\dfrac{7}{3}$

31. $m = \dfrac{3}{4}$, $b = 2$

   (a) $y = \dfrac{3}{4}x + 2$

   (b) $4y = 3x + 8$

   $3x - 4y = -8$

33. $m = 6$, $b = -3$

   (a) $y = 6x - 3$

   (b) $6x - y = 3$

35. $m = -\dfrac{5}{2}$, $b = 6$

   (a) $y = -\dfrac{5}{2}x + 6$

   (b) $2y = -5x + 12$

   $5x + 2y = 12$

37. $m = -2$, $b = \dfrac{1}{2}$

   (a) $y = -2x + \dfrac{1}{2}$

   (b) $2x + y = \dfrac{1}{2}$

   $4x + 2y = 1$

39. $m = \dfrac{1}{2}$, $b = -3$

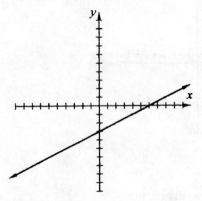

41. $m = -\dfrac{2}{5}$, $b = 3$

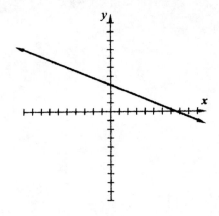

43. $y = \dfrac{3}{4}x - 5$

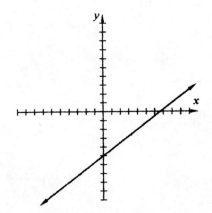

120

45. $y + 3x = 2$

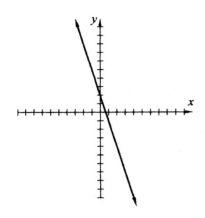

47. $y = 2x$

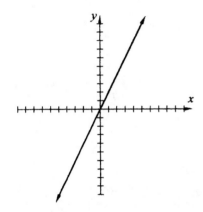

49. $m = \dfrac{13}{5}$

(a) $m_{11} = m = \dfrac{13}{5}$

(b) $m_{\perp} = -\dfrac{1}{m} = -\dfrac{5}{13}$

51. $m = \dfrac{1}{8}$

(a) $m_{11} = m = \dfrac{1}{8}$

(b) $m_{\perp} = -\dfrac{1}{m} = -8$

53. $y = -5x + 1 \Rightarrow m = -5$

(a) $m_{11} = m = -5$

(b) $m_{\perp} = -\dfrac{1}{m} = \dfrac{1}{5}$

55. $y = -\dfrac{1}{2}x - 3 \Rightarrow m = -\dfrac{1}{2}$

(a) $m_{11} = m = -\dfrac{1}{2}$

(b) $m_{\perp} = -\dfrac{1}{m} = 2$

57. $\dfrac{1}{4}x + 3 > \dfrac{2}{3}x + 2$

$3x + 12 \cdot 3 > 4 \cdot 2x + 12 \cdot 2$

$3x + 36 > 8x + 24$

$-5x > -12$

$x < \dfrac{12}{5}$

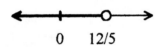

$$\begin{array}{ccc} & & \\ 0 & & 12/5 \end{array}$$

59. $7x - 2(x + 3) \le 4(x - 7)$

$7x - 2x - 6 \le 4x - 28$

$x \le -22$

$$\begin{array}{ccc} & & \\ -22 & 0 & \end{array} \quad 0$$

Putting Your Skills To Work

Use $p = \dfrac{5}{11}d + 15$ in Exercises 1-5.

1. $d = 22$

$p = \dfrac{5}{11}(22) + 15 = 25 \text{ lb / in}^2$

3. $(22, 25)$ and $(44, 35)$

$$m = \frac{35 - 25}{44 - 22} = \frac{10}{22} = \frac{5}{11}$$

5. From any point on the graph, count over 5 in the positive x direction. Then you must count up 11 in the y – direction to get back to the line. This agrees.

7. $(0, 18)$ and $(5, 22)$

$$m = \frac{22 - 18}{5 - 0} = \frac{4}{5}$$

9. When $d = 0$, $p = 18$, so $b = 18$

$$p = \frac{4}{5}d + 18$$

11. The new liquid is more dense because the pressure is greater at the same depth.

Exercises 7.4

Use $y = mx + b$ in Exercises 1-11.

1. $m = 6$, $(-3, 2)$

$$2 = 6(-3) + b \Rightarrow b = 20$$
$$y = 6x + 20$$

3. $m = -2$, $(4, 3)$

$$3 = -2(4) + b \Rightarrow b = 11$$
$$y = -2x + 11$$

5. $m = \frac{2}{3}$, $(-3, -2)$

$$-2 = \frac{2}{3}(-3) + b \Rightarrow b = 0$$
$$y = \frac{2}{3}x$$

7. $m = -3$, $\left(\frac{1}{2}, 2\right)$

$$2 = -3\left(\frac{1}{2}\right) + b \Rightarrow b = \frac{7}{2}$$
$$y = -3x + \frac{7}{2}$$

9. $m = \frac{3}{4}$, $(4, -2)$

$$-2 = \frac{3}{4}(4) + b \Rightarrow b = -5$$
$$y = \frac{3}{4}x - 5$$

11. $m = 0$, $(7, -4)$

$$-4 = 0(7) + b \Rightarrow b = -4$$
$$y = -4$$

Use $m = \dfrac{y_2 - y_1}{x_2 - x_1}$ and $y = mx + b$ in Exercises 13-23.

13. $(1, 5)$ and $(4, 11)$

$$m = \frac{11 - 5}{4 - 1} = \frac{6}{3} = 2$$
$$5 = 2(1) + b \Rightarrow b = 3$$
$$y = 2x + 3$$

15. $(1, -2)$ and $(3, -8)$

$$m = \frac{-8 - (-2)}{3 - 1} = \frac{-6}{2} = -3$$
$$-2 = -3(1) + b \Rightarrow b = 1$$
$$y = -3x + 1$$

17. $(0, -4)$ and $(1, -7)$

$$m = \frac{-7 - (-4)}{1 - 0} = -3$$
$$-4 = -3(0) + b \Rightarrow b = -4$$
$$y = -3x - 4$$

122

19. $(3, 5)$ and $(-1, -15)$

$$m = \frac{-15-5}{-1-3} = \frac{-20}{-4} = 5$$

$$5 = 5(3) + b \Rightarrow b = -10$$

$$y = 5x - 10$$

21. $(1, -3)$ and $(2, 6)$

$$m = \frac{6-(-3)}{2-1} = 9$$

$$-3 = 9(1) + b \Rightarrow b = -12$$

$$y = 9x - 12$$

23. $\left(1, \frac{5}{6}\right)$ and $\left(3, \frac{3}{2}\right)$

$$m = \frac{\frac{3}{2} - \frac{5}{6}}{3-1} = \frac{\frac{4}{6}}{2} = \frac{1}{3}$$

$$\frac{5}{6} = \frac{1}{3}(1) + b \Rightarrow b = \frac{1}{2}$$

$$y = \frac{1}{3}x + \frac{1}{2}$$

25. $b = 1$, $m = -\frac{2}{3}$

$$y = -\frac{2}{3}x + 1$$

27. $b = -4$, $m = \frac{2}{3}$

$$y = \frac{2}{3}x - 4$$

29. $b = 0$, $m = 3$

$$y = 3x$$

31. $b = -2$, $m = 0$

$$y = -2$$

To Think About

33. $(6, -1)$; m is undefined

$$x = 6$$

35. $(-5, 7)$, parallel to y – axis $\Rightarrow m$ is undefined

$$x = -5$$

37. $(0, 2)$, perpendicular to y – axis $\Rightarrow m = 0$

$$y = 2$$

39. $(0, 3)$, parallel to $y = 2x + 1 \Rightarrow m = 2$

$$y = 2x + 3$$

41. $(1, 5)$ parallel to $y = 3x - 1 \Rightarrow m = 3$

$$y = mx + b$$

$$5 = 3(1) + b \Rightarrow b = 2$$

$$y = 3x + 2$$

Cumulative Review Problems

43. $$\frac{1}{8} + \frac{1}{t} = \frac{1}{3}$$

$$3t + 24 = 8t$$

$$24 = 5t$$

$$\frac{24}{5} = t$$

45. Number of microchips shipped

$$44(512) = 22{,}528$$

Percent defective $= \dfrac{231}{22{,}528} \cdot 100\% = 1.03\%$

Exercises 7.5

1. Any convenient point not on the boundary line may be used as a test point.

3. $y > 2x - 2$

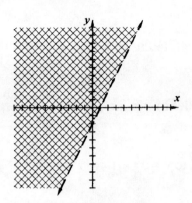

9. $y \geq 3x$

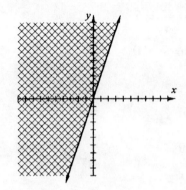

5. $2x - 3y < 6$

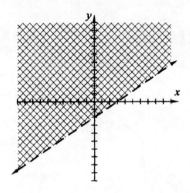

11. $y < -\dfrac{1}{2}x$

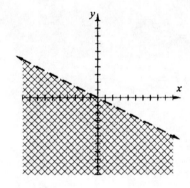

7. $x - y \geq 2$

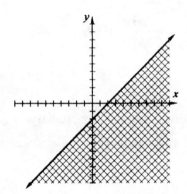

13. $x \leq -3$

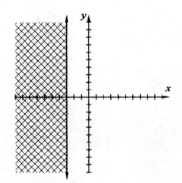

15. $5x - y - 5 \leq 0$

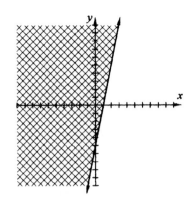

17. $3y \geq -2x$

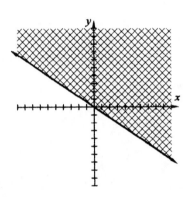

19. $2x > 3 - y$

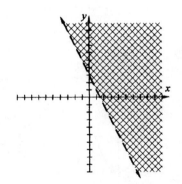

21. $x > -2y$

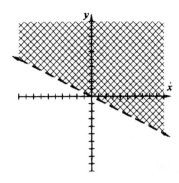

23. $3x + 3 > 5x - 3$

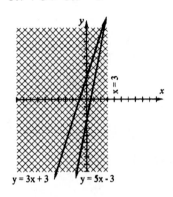

(a) $x < 3$

(b) $3x + 3 > 5x - 3$

$ -2x > -6$

$ x < 3$

(c) $3x + 3$ will be greater than $5x - 3$ for those values of x where the graph of $y = 3x + 3$ lies above the graph of $y = 5x - 3$.

Cumulative Review Problems

25. $\qquad 2x^2 + 7x + 6 = 0$

$2x^2 + 3x + 4x + 6 = 0$

$x(2x + 3) + 2(2x + 3) = 0$

$(x + 2)(2x + 3) = 0$

$x + 2 = 0 \qquad 2x + 3 = 0$

$x = -2 \qquad x = -\dfrac{3}{2}$

125

27.
$x = $ 1st odd integer

$x + 2 = $ 2nd odd integer

$x + 4 = $ 3rd odd integer

$x + x + 2 + x + 4 = 243$

$3x = 237$

$x = 79,$ also

$x + 2 = 81,$ and

$x + 4 = 83$

The consecutive odd integers are 79, 81, 83. 83 is the largest.

Exercises 7.6

1. You can describe a function using a table of values, an algebraic equation, or a graph.

3. The domain of a function is the set of <u>possible</u> values of the <u>independent</u> variable.

5. If a vertical line can intersect the graph more than once, the relation is not a function.

7. $\{(2, 6), (6, 8), (9, 9), (6, 2)\}$

   Domain $= \{2, 6, 9\}$

   Range $= \{2, 6, 8, 9\}$

   Not a function

9. $\left\{\left(\frac{1}{2}, 5\right), \left(\frac{1}{4}, 10\right), \left(\frac{3}{4}, 6\right), \left(\frac{1}{2}, 6\right)\right\}$

   Domain $= \left\{\frac{1}{4}, \frac{1}{2}, \frac{3}{4}\right\}$

   Range $= \{5, 6, 10\}$

   Not a function

11. $\{(10, 1), (1, 10), (8, 3), (3, 8)\}$

    Domain $= \{1, 3, 8, 10\}$

    Range $= \{1, 3, 8, 10\}$

    Function

13. $\{(5.6, 8), (5.8, 6), (6, 5.8), (5, 6)\}$

    Domain $= \{5, 5.6, 5.8, 6\}$

    Range $= \{5.8, 6, 8\}$

    Function

15. $\{(40, 2), (50, 30), (30, 60), (40, 80)\}$

    Domain $= \{30, 40, 50\}$

    Range $= \{2, 30, 60, 80\}$

    Not a function

17. $y = x^2 - 1$

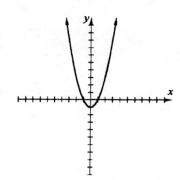

x	y
-2	3
-1	0
0	-1
1	0
2	3

19. $y = \frac{1}{2}x^2$

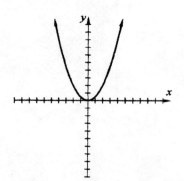

x	y
-3	4.5
-2	2
0	0
2	2
3	4.5

21. $x = y^2 - 3$

126

x	y
1	-2
-2	-1
-3	0
-2	1
1	2

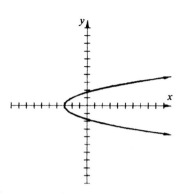

x	y
-3	1/3
-1	1
-1/3	3
1/3	-3
1	-1
3	-1/3

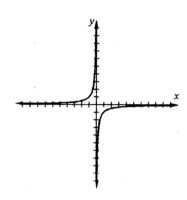

23. $x = 2y^2$

x	y
8	-2
2	-1
0	0
2	1
8	2

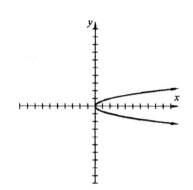

29. $x = (y-2)^2$

x	y
4	0
1	1
0	2
1	3
4	4

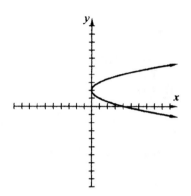

25. $y = -\dfrac{2}{x}$

x	y
-4	1/2
-2	1
-1/2	4
1/2	-4
2	-1
4	-1/2

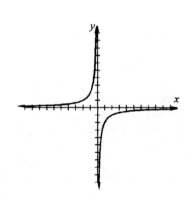

31. $x = \dfrac{2}{y+1}$

x	y
-1/2	-5
-1	-3
-2	-2
2	0
1	1
1/2	3

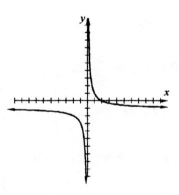

33. Function

27. $y = -\dfrac{1}{x}$

35. Not a function

37. Function

39. Not a function

127

41. $f(x) = 3x - 7$

   (a) $f(1) = 3(1) - 7 = -4$

   (b) $f(-2) = 3(-2) - 7 - 13$

   (c) $f(3) = 3(3) - 7 = 2$

43. $f(x) = 1 - 10x$

   (a) $f(0.5) = 1 - 10(0.5) = -4$

   (b) $f(-0.1) = 1 - 10(-0.1) = 2$

   (c) $f(3) = 1 - 10(3) = -29$

45. $f(x) = x^2 + 2x - 3$

   (a) $f(-1) = (-1)^2 + 2(-1) - 3 = -4$

   (b) $f(0) = (0)^2 + 2(0) - 3 = -3$

   (c) $f(3) = (3)^2 + 2(3) - 3 = 12$

47. $g(x) = 3 - 4x^2$

   (a) $g(0) = 3 - 4(0)^2 = 3$

   (b) $g(2) = 3 - 4(2)^2 = -13$

   (c) $g(-3) = 3 - 4(-3)^2 = -33$

49. $f(x) = \dfrac{6}{x} - 2$

   (a) $f(6) = \dfrac{6}{6} - 2 = -1$

   (b) $f(-3) = \dfrac{6}{-3} - 2 = -4$

   (c) $f(-12) = \dfrac{6}{-12} - 2 = -2\dfrac{1}{2}$

Cumulative Review Problems

51. $\left(-2x^2 - 3x + 8\right) - \left(x^2 + 4x - 3\right)$

   $= \left(-2x^2 - 3x + 8\right) + \left(-x^2 - 4x + 3\right)$

   $= \left(-2 - 1\right)x^2 + \left(-3 - 4x\right) + 8 + 3$

   $= -3x^2 - 7x + 11$

53.
$$\begin{array}{r}
2x + 4 \\
3x - 1 \overline{\smash{\big)}\ 6x^2 + 10x - 5} \\
\underline{6x^2 - 2x} \\
12x - 5 \\
\underline{12x - 4} \\
-1
\end{array}$$

$\left(6x^2 + 10x - 5\right) \div \left(3x - 1\right) = 2x + 4 + \dfrac{-1}{3x - 1}$

Putting Your Skills To Work

1.

Number of Subscribers	Cost per Month	Monthly Income for the Cable Company
10,000	$15.00	$150,000
10,200	$14.75	$150,450
10,400	$14.50	$150,800
10,600	$14.25	$151,050
10,800	$14.00	$151,200
11,000	$13.75	$151,250
11,200	$13.50	$151,200

3.

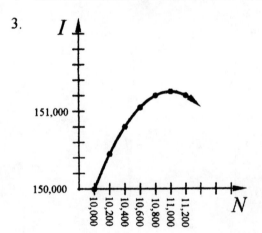

Use $I(x) = (10{,}000 + 200x)(15 - 0.25x)$ in Exercises 5 and 7.

5. From Exercise 4 $I(1) = \$150{,}450$ and $I(2) = \$150{,}800$

   From the table

   $N(1) = 10{,}200$ and $N(2) = 10{,}400$

128

7. From the table, maximum I is \$151,250
$I(s) = [10,000 + 200(5)][15 - 0.25(5)] = 12,150$
The maximum income is obtained when $x = 5$.

Chapter 7 - Review Problems

1. A: $(2, -3)$, B: $(-1, 0)$, C: $(3, 2)$, D: $(-2, -3)$

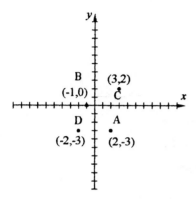

3. $y = 3x - 5$

(a) $(0,)$: $y = 3(0) - 5 = -5$: $(0, -5)$
(b) $(3,)$: $y = 3(3) - 5 = 4$: $(3, 4)$

5. $x = 6$
(a) $(, -1)$: $x = 6$: $(6, -1)$
(b) $(, 3)$: $x = 6$: $(6, 3)$

7. $5y + x = -15$

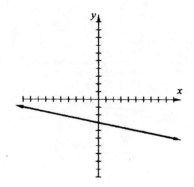

9. $(5, -3)$ and $\left(2, -\dfrac{1}{2}\right)$

$$m = \frac{y_2 - y_1}{x_2 - x_1} = \frac{-\frac{1}{2} - (-3)}{2 - 5} = -\frac{5}{6}$$

11. $m = -\dfrac{1}{2}$, $b = 3$
$y = mx + b$
$y = -\dfrac{1}{2}x + 3$

13. $m = -\dfrac{2}{3}$
$m_\perp = -\dfrac{1}{m} = \dfrac{3}{2}$

15. $y = -\dfrac{1}{2}x + 3$

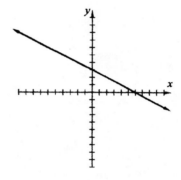

17. $5x + 2y = 20 + 2y$
$5x = 20$
$x = 4$

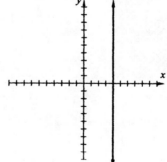

19. $(3, -4)$, $m = -6$

$y = mx + b$

$-4 = -6(3) + b \Rightarrow b = 14$

$y = -6x + 14$

21. $(5, 2)$ and $(-3, 6)$

$m = \dfrac{y_2 - y_1}{x_2 - x_1} = \dfrac{6 - 2}{-3 - 5} = -\dfrac{1}{2}$

$y = mx + b$

$2 = -\dfrac{1}{2}(5) + b \Rightarrow b = \dfrac{9}{2}$

$y = -\dfrac{1}{2}x + \dfrac{9}{2}$

23. $(0, 5)$ and $(-1, 5)$

$m = \dfrac{y_2 - y_1}{x_2 - x_1} = \dfrac{5 - 5}{-1 - 0} = 0$

$y = 5$

25. $b = 1$, $m = -3$

$y = -3x + 1$

27. $y < \dfrac{1}{3}x + 2$

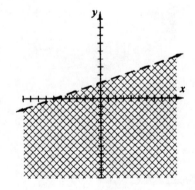

29. $y \geq -2$

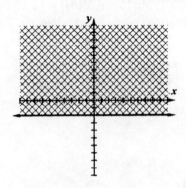

31. $\{(0, 6), (1, 4), (2, 8), (0, 8)\}$

Domain $\{0, 1, 2\}$

Range $\{6, 4, 8\}$

Not a function

33. $\{(3, -7), (-7, 3), (-3, 7), (7, -3)\}$

Domain $\{3, -7, -3, 7\}$

Range $\{-7, 3, 7, -3\}$

Function

35. Not a function

37. $y = x^2 - 5$

x	y
-2	-1
-1	-4
0	-5
1	-4
2	-1

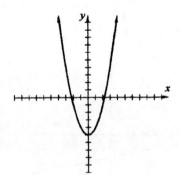

39. $y = (x-3)^2$

x	y
1	4
2	1
3	0
4	1
5	4

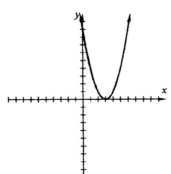

41. $f(x) = 2x^2 - x + 3$

   (a) $f(2) = 2(2)^2 - (2) + 3 = 9$

   (b) $f(-3) = 2(-3)^2 - (-3) + 3 = 24$

43. $h(x) = -5 + 4x^2$

   (a) $h(-3) = -5 + 4(-3)^2 = 31$

   (b) $h(4) = -5 + 4(4)^2 = 59$

45. $f(x) = \dfrac{2}{x+4}$

   (a) $f(-2) = \dfrac{2}{-2+4} = 1$

   (b) $f(6) = \dfrac{2}{6+4} = \dfrac{1}{5}$

Test - Chapter 7

1. $C = (-4, -3)$ $B = (6, 1)$

   $D = (-3, 0)$ $E = (5, -2)$

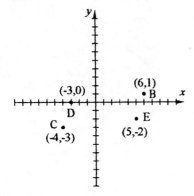

3. $12x - 3y = 6$

   $4x - y = 2$

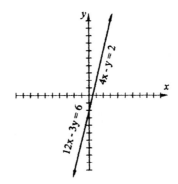

5. $3x + 2y - 5 = 0$

   $2y = -3x + 5$

   $y = -\dfrac{3}{2}x + \dfrac{5}{2}$

   $m = -\dfrac{3}{2}, \quad b = \dfrac{5}{2}$

7. $(4, -2), \quad m = \dfrac{1}{2}$

   $y = mx + b$

   $-2 = \dfrac{1}{2}(4) + b \Rightarrow b = -4$

   $y = \dfrac{1}{2}x - 4 \text{ or } x - 2y = 8$

9. $(2, 5)$ and $(8, 3)$

   $m = \dfrac{y_2 - y_1}{x_2 - x_1} = \dfrac{3-5}{8-2} = -\dfrac{1}{3}$

   $y = mx + b$

   $5 = -\dfrac{1}{3}(2) + b \Rightarrow b = \dfrac{17}{3}$

   $y = -\dfrac{1}{3}x + \dfrac{17}{3} \text{ or } x + 3y = 17$

11. $4y \le 3x$

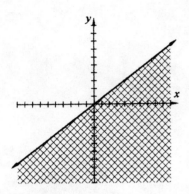

13. $\{(-8, 2), (3, 2), (5, -2)\}$

This is a function because there is only one second element for each first element.

15. $y = 2x^2 - 3$

x	y
-2	5
-1	-1
0	-3
1	-1
2	5

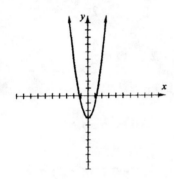

17. $g(x) = \dfrac{3}{x-4}$

   (a) $g(3) = \dfrac{3}{3-4} = -3$

   (b) $g(0) = \dfrac{3}{0-4} = -\dfrac{3}{4}$

Cumulative Tests- Chapters 0 - 7

1. $\left(-3x^3y^4z\right)^3 = -27x^9y^{12}z^3$

3. $2 - 3(4 - x) = x - (3 - x)$
$$2 - 12 + 3x = x - 3 + x$$
$$x = 7$$

5. $x = 2^3 - 6 + 3 - 1$
$$= 8 - 6 + 3 - 1$$
$$= 8 - 2 - 1$$
$$= 5$$

7. $50a^2 - 98b^2$
$$= 2\left(25a^2 - 49b^2\right)$$
$$= 2(5a + 7b)(5a - 7b)$$

9. $\dfrac{x - \frac{1}{x}}{\frac{1}{2} + \frac{1}{2x}} = \dfrac{x - \frac{1}{x}}{\frac{1}{2} + \frac{1}{2x}} \cdot \dfrac{2x}{2x}$
$$= \dfrac{2x^2 - 2}{x + 1}$$
$$= \dfrac{2(x + 1)(x - 1)}{x + 1}$$
$$= 2(x - 1)$$

11. $(7, -4)$ and slope is undefined
$$x = 7$$

13. $(-8, -3)$ and $(11, -3)$
$$m = \dfrac{y_2 - y_1}{x_2 - x_1} = \dfrac{-3 - (-3)}{11 - (-8)} = 0$$

15. $y = \dfrac{2}{3}x - 4$

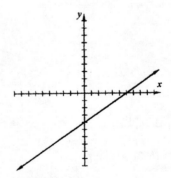

132

17. $2x + 5y \le -10$

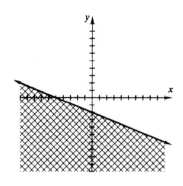

19. $\{(3, 10), (10, -3), (0, -3), (5, 10)\}$

This is a function

Practice Quiz: Sections 7.1 - 7.3

1. Graph the points $(-4, 3)$ and $(5, -1)$.

2. Complete the ordered pair $(-1, \ )$ so that it is a solution to the equation $y = -2x + 5$.

3. Graph $3x - 4y = -12$.

4. What is the slope of the line passing through $(-3, 5)$ and $(2, 1)$?

5. Find the slope of the line $2x - 3y - 5 = 0$.

Practice Quiz: Sections 7.4 - 7.6

1. Find the equation of a line with slope -2 that passes through the point $(-3, 9)$.

2. Find the equation of a line that passes through $(-1, -8)$ and $(3, 4)$.

3. Graph $2x - 5y \le 10$.

4. Determine if $\{(-3, 2), (3, 7), (0, 2)\}$ is a function.

5. If $f(x) = 2x^2 - 3x$, find $f(-1)$.

Answers to Practice Quiz

Sections 7.1 - 7.3

1.

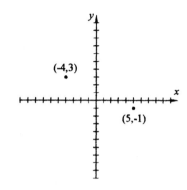

2. $(-1, 7)$

3.

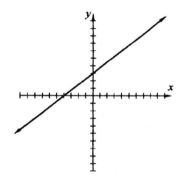

4. $\dfrac{-4}{5}$

5. $\dfrac{2}{3}$

Answers to Practice Quiz

Sections 7.4 - 7.6

1. $y = -2x + 3$

2. $y = 3x - 5$

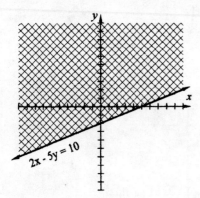

2x - 5y = 10

4. Yes

5. 5

1. $4x + 2y = -8$
   $-2x + 3y = 12$

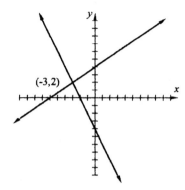

3. $3x + y = -11$ (1)
   $4x - 3y = -6$ (2)
   Solve (1) for y
   $y = -3x - 11$ (3)
   Substitute $-3x - 11$ for y in (2)
   $4x - 3(-3x - 11) = -6$
   $\quad 4x + 9x + 33 = -6$
   $\quad\quad\quad\quad 13x = -39$
   $\quad\quad\quad\quad\quad x = -3$
   Substitute -3 for x in (3)
   $y = -3(-3) - 11 = -2$

5. $5x - 3y = 10$ (1)
   $4x + 7y = 8$ (2)
   Multiply (1) by 7 and (2) by 3
   $35x - 21y = 70$
   $\underline{12x + 21y = 24}$
   $47x \quad\quad = 94$
   $\quad\quad x = 2$

5. continued

     Substitute 2 for x in (2)
     $4(2) + 7y = 8$
          $7y = 0$
           $y = 0$

7. $0.8x - 0.3y = 0.7$ (1)
   $1.2x + 0.6y = 4.2$ (2)
   Multiply (1) by 20 and (2) by 10
   $16x - 6y = 14$
   $\underline{12x + 6y = 42}$
   $28x \quad\quad = 56$
   $\quad\quad x = 2$
   Substitute 2 for x in (2)
   $1.2(2) + 0.6y = 4.2$
   $\quad\quad\quad 0.6y = 1.8$
   $\quad\quad\quad\quad y = 3$

9. If a system has no solutions, the graphs of the equations will be lines that are parallel.

11. $x + 8y = 7$ (1)
    $5x - 2y = 14$ (2)
    Multiply (2) by 4
    $x + 8y = 7$
    $\underline{20x - 8y = 56}$
    $21x \quad\quad = 63$
    $\quad\quad x = 3$
    Substitute 3 for x in (1)
    $3 + 8y = 7$
    $\quad 8y = 4$
    $\quad\quad y = \dfrac{1}{2}$

13. $2x - 3y = 7$ (1)

$3x - 5y = 11$ (2)

Add 3 times (1) to -2 times (2)

$6x - 9y = 21$

$\underline{-6x + 10y = -22}$

$y = -1$

Substitute -1 for y in (1)

$2x - 3(-1) = 7$

$2x = 4$

$x = 2$

Exercises 8.1

1. $x + y = 8$ (5,3)

$x - y = 2$

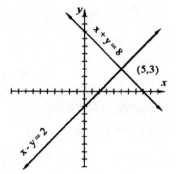

3. $2x + y = 6$ $(4, -2)$

$x - 2y = 8$

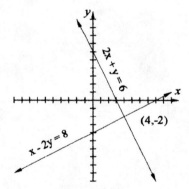

5. $2x + y = 6$ (1,4)

$-2x + y = 2$

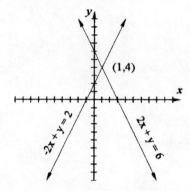

7. $-2x + y - 3 = 0$ $(-1,1)$

$4x + y + 3 = 0$

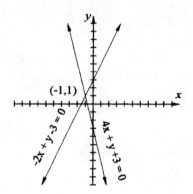

9. $3x - 2y = -18$ $(-2,6)$

$2x + 3y = 14$

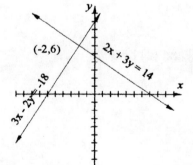

136

11. $y = \dfrac{2}{3}x - 3$ $(3, -1)$

$y = -2x + 5$

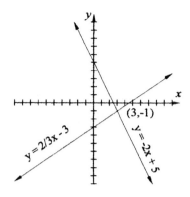

13. $4x - 6y = 8$ Infinite number of solutions.

$-2x + 3y = -4$ Dependent system.

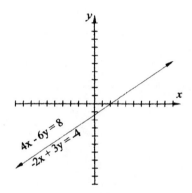

15. $2y + x - 6 = 0$ No solution

$y + \dfrac{1}{2}x = 4$ Inconsistent system.

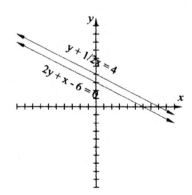

17. $\dfrac{1}{7}x - \dfrac{3}{7}y = 1$ $\left(-\dfrac{19}{5}, -\dfrac{18}{5}\right)$

$\dfrac{1}{2}x - \dfrac{1}{4}y = -1$

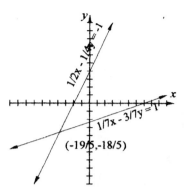

19. The lines $y = 2x - 5$ and $y = 2x + 6$ are parallel with y – intercepts at -5 and 6 respectively. There is no solution.

21. If two lines have different slopes you can conclude they <u>intersect</u> and have <u>one</u> solution.

23. If two lines have different slopes but the same y – intercept, the lines intersect at the y – intercept which is the solution to the system.

<u>Cumulative Review Problems</u>

25. $3dx - 5y = 2(dx + 8)$

$3dx - 5y = 2dx + 16$

$dx = 5y + 16$

$d = \dfrac{5y + 16}{x}$

27. $25x^2 - 60x + 36 = (5x - 6)^2$

1. $y = 4x - 3$ (1)

    $3x - y = 2$ (2)

    Substitute $4x - 3$ for y in (2).

    $3x - (4x - 3) = 2$

    $3x - 4x + 3 = 2$

    $-x + 3 = 2$

    $-x = -1$

    $x = 1$

    Substitute 1 for x in (1)

    $y = 4(1) - 3 = 1$

    $(1, 1)$

    Check: $3(1) - (1) \overset{?}{=} 2$

               $2 = 2$

3. $4x + 3y = 9$ (1)

    $x - 3y = 6$ (2)

    Solve (2) for x

    $x = 3y + 6$ (3)

    Substitute $3y + 6$ for x in (1)

    $4(3y + 6) + 3y = 9$

    $12y + 24 + 3y = 9$

    $15y + 24 = 9$

    $15y = -15$

    $y = -1$

    Substitute -1 for y in (3)

    $x = 3(-1) + 6 = 3$

    $(3, -1)$

    Check: $4(3) + 3(-1) \overset{?}{=} 9$

               $9 = 9$

5. $2x + y = 4$ (1)

    $2x - y = 0$ (2)

    Solve (1) for y

    $y = -2x + 4$ (3)

    Substitute $-2x + 4$ for y in (2)

    $2x - (-2x + 4) = 0$

    $2x + 2x - 4 = 0$

    $4x - 4 = 0$

    $4x = 4$

    $x = 1$

    Substitute 1 for x in (3)

    $y = -2(1) + 4 = 2$

    $(1, 2)$

    Check: $2(1) - 2 \overset{?}{=} 0$

               $0 = 0$

7. $3x - 4y = -8$ (1)

    $y = 2x$ (2)

    Substitute $2x$ for y in (1)

    $3x - 4(2x) = -8$

    $3x - 8x = -8$

    $-5x = -8$

    $x = \dfrac{8}{5}$

    Substitute $\dfrac{8}{5}$ for x in (2)

    $y = 2\left(\dfrac{8}{5}\right) = \dfrac{16}{5}$

    $\left(\dfrac{8}{5}, \dfrac{16}{5}\right)$

    Check: $3\left(\dfrac{8}{5}\right) - 4\left(\dfrac{16}{5}\right) \overset{?}{=} -8$

                    $-8 = -8$

9. $x + 3y = 2$ (1)

$2x + 3y = 7$ (2)

Solve (1) for x

$x = -3y + 2$ (3)

Substitute $-3y + 2$ for x in (2)

$2(-3y + 2) + 3y = 7$

$-6y + 4 + 3y = 7$

$-3y + 4 = 7$

$-3y = 3$

$y = -1$

Substitute -1 for y in (3)

$x = -3(-1) + 2 = 5$

$(5, -1)$

Check: $2(5) + 3(-1) \overset{?}{=} 7$

$7 = 7$

11. $-x + 2y = -3$ (1)

$-2x + 3y = -6$ (2)

Solve (1) for x

$-x = -2y - 3$

$x = 2y + 3$ (3)

Substitute $2y + 3$ for x in (2)

$-2(2y + 3) + 3y = -6$

$-4y - 6 + 3y = -6$

$-y - 6 = -6$

$-y = 0$

$y = 0$

Substitute 0 for y in (3)

$x = 2(0) + 3 = 3$

$(3, 0)$

Check: $-2(3) + 3(0) \overset{?}{=} -6$

$-6 = -6$

13. $3a - 5b = 2$ (1)

$3a + b = 32$ (2)

Solve (2) for b

$b = -3a + 32$ (3)

Substitute $-3a + 32$ for b in (1)

$3a - 5(-3a + 32) = 2$

$3a + 15a - 160 = 2$

$18a - 160 = 2$

$18a = 162$

$a = 9$

Substitute 9 for a in (3)

$b = -3(9) + 32 = 5$

$(9, 5)$

Check: $3(9) - 5(5) \overset{?}{=} 2$

$2 = 2$

15. $3x - 4y = 6$ (1)

$-2x - y = -4$ (2)

Solve (2) for y

$-y = 2x - 4$

$y = -2x + 4$ (3)

Substitute $-2x + 4$ for y in (1)

$3x - 4(-2x + 4) = 6$

$3x + 8x - 16 = 6$

$11x - 16 = 6$

$11x = 22$

$x = 2$

Substitute 2 for x in (3)

$y = -2(2) + 4 = 0$

$(2, 0)$

Check: $3(2) - 4(0) \overset{?}{=} 6$

$6 = 6$

17. $5c + 8d + 1 = 0$ (1)

$3c + d - 7 = 0$ (2)

Solve (2) for d

$d = -3c + 7$ (3)

Substitute $-3c + 7$ for d in (1)

$5c + 8(-3c + 7) + 1 = 0$

$5c - 24c + 56 + 1 = 0$

$-19c + 57 = 0$

$57 = 19c$

$3 = c$

Substitute 3 for c in (3)

$d = -3(3) + 7 = -2$

$(3, -2)$

Check: $5(3) + 8(-2) + 1 \overset{?}{=} 0$

$0 = 0$

19. $3x - y - 9 = 0$ (1)

$8x + 5y - 1 = 0$ (2)

Solve (1) for y

$3x - 9 = y$ (3)

Substitute $3x - 9$ for y in (2)

$8x + 5(3x - 9) - 1 = 0$

$8x + 15x - 45 - 1 = 0$

$23x - 46 = 0$

$x = 2$

Substitute 2 for x in (3)

$y = 3(2) - 9 = -3$

$(2, -3)$

Check: $8(2) + 5(-3) - 1 \overset{?}{=} 0$

$0 = 0$

21. $\dfrac{5}{3}x + \dfrac{1}{3}y = -3$ (1)

$-2x + 3y = 24$ (2)

Multiply (1) by 3

$5x + y = -9$ (3)

Solve (3) for y

$y = -5x - 9$ (4)

Substitute $-5x - 9$ for y in (2)

$-2x + 3(-5x - 9) = 24$

$-2x - 15x - 27 = 24$

$-17x = 51$

$x = -3$

Substitute -3 for x in (4)

$y = -5(-3) - 9 = 6$

$(-3, 6)$

Check: $-2(-3) + 3(6) \overset{?}{=} 24$

$24 = 24$

23. $3x - 2y = -11$ (1)

$x - \dfrac{1}{3}y = \dfrac{-10}{3}$ (2)

Multiply (2) by 3

$3x - y = -10$

Solve for x

$x = \dfrac{y - 10}{3}$ (3)

Substitute $\dfrac{y - 10}{3}$ for x in (1)

$\dfrac{3(y - 10)}{3} - 2y = -11$

$y - 10 - 2y = -11$

$-y = -1$

$y = 1$

23. continued

Substitute 1 for y in (3)

$$x = \frac{1-10}{3} = -3$$

$(-3, 1)$

Check: $3(-3) - 2(1) \overset{?}{=} -11$

$$-11 = -11$$

25. $\frac{3}{5}x - \frac{2}{5}y = 5$ (1)

    $x + 4y = -1$ (2)

Multiply (1) by 5

$3x - 2y = 25$

Solve for x

$$x = \frac{2y + 25}{3} \qquad (3)$$

Substitute $\frac{2y + 25}{3}$ for x in (2)

$$\frac{2y + 25}{3} + 4y = -1$$

$$2y + 25 + 12y = -3$$

$$14y = -28$$

$$y = -2$$

Substitute -2 for y in (3)

$$x = \frac{2(-2) + 25}{3} = 7$$

$(7, -2)$

Check: $\frac{3}{5}(7) - \frac{2}{5}(-2) \overset{?}{=} 5$

$$5 = 5$$

27. $-\frac{1}{2}x + \frac{1}{6}y = 2$ (1)

    $3x - 2y = -15$ (2)

Multiply (1) by 6

$-3x + y = 12$

Solve for y

$y = 3x + 12$ (3)

27. continued

Substitute $3x + 12$ for y in (2)

$$3x - 2(3x + 12) = -15$$

$$3x - 6x - 24 = -15$$

$$-3x = 9$$

$$x = -3$$

Substitute -3 for x in (3)

$$y = 3(-3) + 12 = 3$$

$(-3, 3)$

Check: $-\frac{1}{2}(-3) + \frac{1}{6}(3) \overset{?}{=} 2$

$$2 = 2$$

29. $3(x + y) = 10 - y$ (1)

    $x - 2 = 3(y - 1)$ (2)

Solve (2) for x

$x = 3y - 1$ (3)

Substitute $3y - 1$ for x in (1)

$$3(3y - 1 + y) = 10 - y$$

$$12y - 3 = 10 - y$$

$$13y = 13$$

$$y = 1$$

Substitute 1 for y in (3)

$$x = 3(1) - 1 = 2$$

$(2, 1)$

Check: $3(2 + 1) \overset{?}{=} 10 - 1$

$$9 = 9$$

31. $\frac{1}{4}x - 3y = 0$ (1)

    $\frac{3}{4}x + 6y = \frac{15}{4}$ (2)

Multiply (1) and (2) by 4

    $x - 12y = 0$ (3)

    $3x + 24y = 15$ (4)

Solve (3) for x

    $x = 12y$ (5)

31. continued

Substitute $12y$ for x in (2)

$$\frac{3}{4}(12y) + 6y = \frac{15}{4}$$

$$9y + 6y = \frac{15}{4}$$

$$15y = \frac{15}{4}$$

$$y = \frac{1}{4}$$

Substitute $\frac{1}{4}$ for y in (5)

$$x = 12\left(\frac{1}{4}\right)$$

$$x = 3$$

$$\left(3, \frac{1}{4}\right)$$

Check: $\dfrac{3}{4}(3) + 6\left(\dfrac{1}{4}\right) \overset{?}{=} \dfrac{15}{4}$

$$\frac{15}{4} = \frac{15}{4}$$

33. $2.9864x + \quad y = 9.8632 \qquad (1)$
$1.8926x + 4.7618y = 10.7891 \qquad (2)$
Solve (1) for y
$y = -2.9864x + 9.8632 \qquad (3)$
Substitute $-2.9864x + 9.8632$ for y in (2)
$1.8926x + 4.7618(-2.9864x + 9.8632) = 10.7891$

$$-12.32804x = -36.177486$$
$$x = 2.9346$$

Substitute 2.9346 for x in (3)
$y = -2.9864(2.9346) + 9.8632$
$\quad = 1.0993$
$(2.9346, 1.0993)$

35. You would need 3 (7) equations to solve for 3(7) unknowns. Substitution can reduce the system to one equation with one unknown. Then substituting back successively gives another unknown, thus requiring one equation for each unknown.

37. The solution to a system must satisfy <u>both</u> equations.

39. If the graphs are parallel, the system has no solutions because there are no points in common.

<u>Cumulative Review Problems</u>

41. $3x - y < 2$

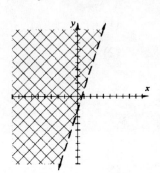

43.
$$x = \text{number of dimes}$$
$$154 - x = \text{number of quarters}$$
$$0.1x + 0.25(154 - x) = 34$$
$$10x + 25(154 - x) = 3400$$
$$10x + 3850 - 25x = 3400$$
$$-15x = -450$$
$$x = 30, \text{ and}$$
$$154 - x = 124$$

He has 30 dimes and 124 quarters.

<u>Exercises 8.3</u>

1. (a) $3x + 2y = 5 \qquad (1)$
$\quad\; 5x - \;\, y = 3 \qquad (2)$
Multiply (2) by 2 to eliminate y

(b) $2x - 9y = 1 \qquad (1)$
$\quad 2x + 3y = 2 \qquad (2)$
Multiply either (1) or (2) by -1
to eliminate x

142

(c) $4x - 3y = 10$ (1)

$5x + 4y = 0$ (2)

Multiply (1) by 4 and (2) by 3
to eliminate y

3. $2x + y = 2$ (1)

$\underline{4x - y = -8}$ (2)

$6x = -6$

$x = -1$

Substitute -1 for x in (1)

$2(-1) + y = 2$

$y = 4$

$(-1, 4)$

Check: $4(-1) - 4 = -8$

$-8 = -8$

5. $2x + y = 4$ (1)

$3x - 2y = -1$ (2)

Multiply (1) by 2

$4x + 2y = 8$

$\underline{3x - 2y = -1}$

$7x = 7$

$x = 1$

Substitute 1 for x in (1)

$2(1) + y = 4$

$y = 2$

$(1, 2)$

Check: $3(1) - 2(2) \overset{?}{=} -1$

$-1 = -1$

7. $x + y = 5$ (1)

$\underline{2x - y = -5}$ (2)

$3x = 0$

$x = 0$

Substitute 0 for x in (1)

$0 + y = 5$

$y = 5$

$(0, 5)$

Check: $2(0) - 5 \overset{?}{=} -5$

$-5 = -5$

9. $2x + 3y = 6$ (1)

$-3x + y = 13$ (2)

Multiply (2) by -3

$2x + 3y = 6$

$\underline{9x - 3y = -39}$

$11x = -33$

$x = -3$

Substitute -3 for x in (2)

$-3(-3) + y = 13$

$y = 4$

$(-3, 4)$

Check: $2(-3) + 3(4) \overset{?}{=} 6$

$6 = 6$

11. $2x + 5y = 11$ (1)

$3x + 8y = 16$ (2)

Multiply (1) by 3 and (2) by -2

$6x + 15y = 33$

$\underline{-6x - 16y = -32}$

$-y = 1$

$y = -1$

Substitute -1 for y in (1)

$2x + 5(-1) = 11$

$\qquad 2x = 16$

$\qquad x = 8$

$(8, -1)$

Check: $3(8) + 8(-1) \overset{?}{=} 16$

$\qquad\qquad\qquad 16 = 16$

13. $\quad a + 3b = 0 \qquad$ (1)

$\quad -3a - 10b = -2 \qquad$ (2)

Multiply (1) by 3

$\quad 3a + 9b = 0$

$\underline{-3a - 10b = -2}$

$\quad -b = -2$

$\quad b = 2$

Substitute 2 for b in (1)

$a + 3(2) = 0$

$\qquad a = -6$

$(-6, 2)$

Check: $-3(-6) - 10(2) \overset{?}{=} -2$

$\qquad\qquad\qquad -2 = -2$

15. $4x + 9y = 0 \qquad$ (1)

$\quad 8x - 5y = -23 \qquad$ (2)

Multiply (1) by -2

$-8x - 18y = 0$

$\underline{8x - 5y = -23}$

$-23y = -23$

$\quad y = 1$

Substitute 1 for y in (1)

$4x + 9(1) = 0$

$\qquad x = -\dfrac{9}{4}$

$\left(-\dfrac{9}{4}, 1\right)$

Check: $8\left(-\dfrac{9}{4}\right) - 5(1) \overset{?}{=} -23$

$\qquad\qquad\qquad -23 = -23$

17. $9x - 2y = 14 \qquad$ (1)

$\quad 4x + 3y = 14 \qquad$ (2)

Multiply (1) by 3 and (2) by 2

$27x - 6y = 42$

$\underline{8x + 6y = 28}$

$35x \qquad = 70$

$\quad x = 2$

Substitute 2 for x in (1)

$9(2) - 2y = 14$

$\quad -2y = -4$

$\qquad y = 2$

$(2, 2)$

Check: $4(2) + 3(2) \overset{?}{=} 14$

$\qquad\qquad\qquad 14 = 14$

19. $-5x + 4y = -13 \qquad$ (1)

$\quad 11x + 6y = -1 \qquad$ (2)

Multiply (1) by 3 and (2) by -2

$-15x + 12y = -39$

$\underline{-22x - 12y = 2}$

$-37x \qquad = -37$

$\quad x = 1$

Substitute 1 for x in (1)

$-5(1) + 4y = -13$

$\qquad 4y = -8$

$\qquad y = -2$

$(1, -2)$

Check: $11(1) + 6(-2) \overset{?}{=} -1$

$\qquad\qquad\qquad -1 = -1$

21. $5x - 2y = 24$ (1)

 $4x - 5y = 9$ (2)

 Multiply (1) by 4 and (2) by -5

 $20x - 8y = 96$

 $\underline{-20x + 25y = -45}$

 $17y = 51$

 $y = 3$

 Substitute 3 for y in (2)

 $4x - 5(3) = 9$

 $4x = 24$

 $x = 6$

 $(6, 3)$

 Check: $5(6) - 2(3) \overset{?}{=} 24$

 $24 = 24$

23. $x + \dfrac{5}{4}y = \dfrac{9}{4}$ (1)

 $\dfrac{2}{5}x - y = \dfrac{3}{5}$ (2)

 Multiply (1) by 4 and (2) by 5

 $4x + 5y = 9$

 $\underline{2x - 5y = 3}$

 $6x = 12$

 $x = 2$

 Substitute 2 for x in (1)

 $2 + \dfrac{5}{4}y = \dfrac{9}{4}$

 $\dfrac{5}{4}y = \dfrac{1}{4}$

 $y = \dfrac{1}{5}$

 $\left(2, \dfrac{1}{5}\right)$

 Check: $\dfrac{2}{5}(2) - \dfrac{1}{5} \overset{?}{=} \dfrac{3}{5}$

 $\dfrac{3}{5} = \dfrac{3}{5}$

25. $\dfrac{4}{3}x + \dfrac{4}{5}y = 0$ (1)

 $-2x + 6y = 36$ (2)

 Multiply (1) by 15 and (2) by -2

 $20y + 12y = 0$

 $\underline{4x - 12y = -72}$

 $24x = -72$

 $x = -3$

 Substitute -3 for x in (1)

 $\dfrac{4}{3}(-3) + \dfrac{4}{5}y = 0$

 $\dfrac{4}{5}y = 4$

 $y = 5$

 $(-3, 5)$

 Check: $-2(-3) + 6(5) \overset{?}{=} 36$

 $36 = 36$

27. $\dfrac{5}{6}x + y = -\dfrac{1}{3}$ (1)

 $-8x + 9y = 28$ (2)

 Multiply (1) by 18 and (2) by -2

 $15x + 18y = -6$

 $\underline{16x - 18y = -56}$

 $31x = -62$

 $x = -2$

 Substitute -2 for x in (2)

 $-8(-2) + 9y = 28$

 $9y = 12$

 $y = \dfrac{4}{3}$

 $\left(-2, \dfrac{4}{3}\right)$

 Check: $\dfrac{5}{6}(-2) + \dfrac{4}{3} \overset{?}{=} -\dfrac{1}{3}$

 $-\dfrac{1}{3} = -\dfrac{1}{3}$

29. $\frac{1}{5}x - \frac{1}{5}y = 1$ (1)

$\frac{1}{6}x + \frac{1}{3}y = -\frac{1}{6}$ (2)

Multiply (1) by 5 and (2) by -6

$\begin{array}{r} x - y = 5 \\ -x - 2y = 1 \\ \hline -3y = 6 \\ y = -2 \end{array}$

Substitute -2 for y in (1)

$\frac{1}{5}x - \frac{1}{5}(-2) = 1$

$\frac{1}{5}x = \frac{3}{5}$

$x = 3$

$(3, -2)$

Check: $\frac{1}{6}(3) + \frac{1}{3}(-2) \overset{?}{=} -\frac{1}{6}$

$-\frac{1}{6} = -\frac{1}{6}$

31. (a) $0.5x - 0.3y = 0.1$ (1)

      $5x + 3y = 6$ (2)

Multiply (1) by 10

$5x - 3y = 1$

$5x + 3y = 6$

(b) $0.08x + y = 0.05$ (1)

      $2x - 0.1y = 3$ (2)

Multiply (1) by 100 and (2) by 10

$8x + 100y = 5$

$20x - y = 30$

(c) $4x + 0.5y = 9$ (1)

      $0.2x - 0.05y = 1$ (2)

Multiply (1) by 10 and (2) by 100

$40x + 5y = 90$

$20x - 5y = 100$

33. $0.5x - 0.2y = 0.5$ (1)

    $0.4x + 0.7y = 0.4$ (2)

Multiply (1) by 10 and (2) by 10

$5x - 2y = 5$ (3)

$4x + 7y = 4$ (4)

Multiply (3) by 7 and (4) by 2

$\begin{array}{r} 35x - 14y = 35 \\ 8x + 14y = 8 \\ \hline 43x = 43 \\ x = 1 \end{array}$

Substitute 1 for x in (3)

$5(1) - 2y = 5$

$y = 0$

$(1, 0)$

Check: $0.5(1) - 0.2(0) \overset{?}{=} 0.5$

$0.5 = 0.5$

35. $0.04x - 0.03y = 0.05$ (1)

    $0.05x + 0.08y = -0.76$ (2)

Multiply (1) and (2) by 100

$4x - 3y = 5$ (3)

$5x + 8y = -76$ (4)

Multiply (3) by 5 and (4) by -4

$\begin{array}{r} 20x - 15y = 25 \\ -20x - 32y = 304 \\ \hline -47y = 329 \\ y = -7 \end{array}$

Substitute -7 for y in (3)

$4x - 3(-7) = 5$

$4x = -16$

$x = -4$

$(-4, -7)$

Check: $0.04(-4) - 0.03(-7) \overset{?}{=} 0.05$

$0.05 = 0.05$

37. $-0.6x - 0.08y = -4$ (1)

     $3x + 2y = 4$ (2)

Multiply (1) by 100 and (2) by 20

$-60x - 8y = -400$

$\underline{60x + 40y = 80}$

     $32y = -320$

     $y = -10$

Substitute -10 for y in (2)

$3x + 2(-10) = 4$

   $3x = 24$

    $x = 8$

$(8, -10)$

Check: $-0.6(8) - 0.08(-10) \overset{?}{=} -4$

                  $-4 = -4$

39. $3x + 5y = 11$ (1)

   $291x + 243y = 1551$ (2)

Multiply (1) by -97

$-291x - 485y = -1067$

$\underline{291x + 243y = 1551}$

    $-242y = 484$

       $y = -2$

Substitute -2 for y in (1)

$3x + 5(-2) = 11$

    $3x = 21$

     $x = 7$

$(7, -2)$

Check: $291(7) + 243(-2) \overset{?}{=} 1551$

            $1551 = 1551$

To Think About

41. $5(x - y) = -7 - 3y$ (1)

   $3(x + 2) = y + 1$ (2)

Expand the left side of (1) and (2) and rearrange.

$5x - 2y = -7$ (3)

$3x - y = -5$ (4)

41. continued

Multiply (4) by -2

$5x - 2y = -7$

$\underline{-6x + 2y = 10}$

$-x = 3$

    $x = -3$

Substitute -3 for x in (4)

$3(-3) - y = -5$

    $-y = 4$

     $y = -4$

$(-3, -4)$

43. $4(x - y) = 8 - y$ (1)

     $y = 2(7 - x)$ (2)

Expand and rearrange

$4x - 3y = 8$ (3)

$2x + y = 14$ (4)

Multiply (4) by 3

$4x - 3y = 8$

$\underline{6x + 3y = 42}$

$10x = 50$

   $x = 5$

Substitute 5 for x in (2)

$y = 2(7 - 5)$

$y = 4$

$(5, 4)$

45. $-\dfrac{x}{a} + \dfrac{y}{b} = -2$ (1)

   $\dfrac{x}{a} + \dfrac{y}{b} = 6$ (2)

   $\underline{}$

   $\dfrac{2y}{b} = 4$

   $y = 2b$

45. continued

Substitute $2b$ for y in (2)

$$\frac{x}{a} + \frac{2b}{b} = 6$$

$$\frac{x}{a} = 4$$

$$x = 4a$$

$$(4a, 2b)$$

Cumulative Review Problems

47. $x =$ the larger number

$y =$ the smaller number

$$x - y = 11 \qquad (1)$$
$$2x + 3y = 7 \qquad (2)$$

Multiply (1) by 3

$$3x - 3y = 33$$
$$\underline{2x + 3y = 7}$$
$$5x \qquad = 40$$
$$x = 8$$

Substitute 8 for x in (1)

$$8 - y = 11$$
$$y = -3$$

The numbers are 8 and -3

49. $6x^2 - 25x + 25$

$$= 6x^2 - 15x - 10x + 25$$
$$= 3x(2x - 5) - 5(2x - 5)$$
$$= (3x - 5)(2x - 5)$$

Putting Your Skills To Work

1. $x + 5y = 6.50$

$2x + 9y = 11.90$

3. $3x + 10y = 20.30 \qquad (1)$

$2x + 11y = 19.60 \qquad (2)$

Multiply (1) by 2 and (2) by -3

$$6x + 20y = 40.60$$
$$\underline{-6x - 33y = -58.80}$$
$$-13y = -18.20$$
$$y = 1.40$$

Substitute 1.40 for y in (2)

$$2x + 11(1.40) = 19.60$$
$$2x = 4.20$$
$$x = 2.10$$

Oil costs $2.10 per quart

Gasoline costs $1.40 per gallon

5. $x =$ number of women's sweaters

$y =$ number of men's sweaters

$$4x + 3y = 340 \qquad (1)$$
$$0.8x + 1.6y = 108 \qquad (2)$$

Multiply (1) by -2 and (2) by 10

$$-8x - 6y = -680$$
$$\underline{8x + 16y = 1080}$$
$$10y = 400$$
$$y = 40$$

Substitute 40 for y in (1)

$$4x + 3(40) = 340$$
$$4x = 220$$
$$x = 55$$

Women's sweaters: $55, men's sweaters: $40

Exercises 8.4

1. If there is no solution to a system of linear equations, the graphs of the equations are parallel lines. Solving the system algebraically, you will obtain an equation that is inconsistent with known facts.

3. If there is exactly one solution, the graphs of the equations <u>intersect</u>. This system is said to be <u>independent</u> and <u>consistent</u>.

5. $x - 4y = 6$ (1)

$\underline{-x + 2y = 4}$ (2)

   $-2y = 10$

      $y = -5$

Substitute -5 for y in (1)

$x - 4(-5) = 6$

      $x = -14$

$(-14, -5)$

7. $4x - 6y = 10$ (1)

$-10x + 15y = -25$ (2)

Multiply (1) by 5 and (2) by 2

   $20x - 30y = 50$

$\underline{-20x + 30y = -50}$

       $0 = 0$

Infinite number of solutions

9. $-2x - 3y = 15$ (1)

  $5x + 2y = 1$ (2)

Multiply (1) by 2 and (2) by 3

$-4x - 6y = 30$

$\underline{15x + 6y = 3}$

$11x \quad\quad = 33$

     $x = 3$

Substitute 3 for x in (2)

$5(3) + 2y = 1$

     $2y = -14$

      $y = -7$

$(3, -7)$

11. $5x + 10y = 15$ (1)

$2x + 4y = -1$ (2)

Multiply (1) by 2 and (2) by -5

  $10x + 20y = 30$

$\underline{-10x - 20y = 5}$

       $0 = 35$

No solution

13. $5x - 3y = 9$ (1)

$7x + 2y = -6$ (2)

Multiply (1) by 2 and (2) by 3

$10x - 6y = 18$

$\underline{21x + 6y = -18}$

$31x \quad\quad = 0$

     $x = 0$

Substitute 0 for x in (2)

$7(0) + 2y = -6$

      $y = -3$

$(0, -3)$

15. $4x - 3y = 3$ (1)

$6x + 5y = 33$ (2)

Multiply (1) by 5 and (2) by 3

$20x - 15y = 15$

$\underline{18x + 15y = 99}$

$38x \quad\quad = 114$

     $x = 3$

Substitute 3 for x in (1)

$4(3) - 3y = 3$

    $-3y = -9$

      $y = 3$

$(3, 3)$

17. $0.3x - 0.5y = 0.4$ (1)

$0.6x + 1.0y = 2.8$ (2)

Multiply (1) by 20 and (2) by -10

$6x - 10y = 8$

$\underline{-6x - 10y = -28}$

$-20y = -20$

$y = 1$

Substitute 1 for y in (1)

$0.3x - 0.5(1) = 0.4$

$0.3x = 0.9$

$x = 3$

$(3, 1)$

19. $4x = 3(5 - y)$ (1)

$2(x - 2y) = 1 + y$ (2)

Expand and rearrange

$4x + 3y = 15$ (3)

$2x - 5y = 1$ (4)

Multiply (4) by -2

$4x + 3y = 15$

$\underline{-4x + 10y = -2}$

$13y = 13$

$y = 1$

Substitute 1 for y in (1)

$4x = 3(5 - 1)$

$x = 3$

$(3, 1)$

21. $3(b - 1) = 2(a + b)$ (1)

$4(a - b) = -b + 1$ (2)

Expand and rearrange

$2a - b = -3$ (3)

$4a - 3b = 1$ (4)

21. continued

Multiply (3) by -2

$-4a + 2b = 6$

$\underline{4a - 3b = 1}$

$-b = 7$

$b = -7$

Substitute -7 for b in (3)

$2a - (-7) = -3$

$2a = -10$

$a = -5$

$(-5, -7)$

23. $26x - 39y = 208$ (1)

$51x + 68y = 119$ (2)

Multiply (1) by 51 and (2) by -26

$1326x - 1989y = 10{,}608$

$\underline{-1326x - 1768y = -3094}$

$-3757y = 7{,}514$

$y = -2$

Substitute -2 for y in (1)

$26x - 39(-2) = 208$

$26x = 130$

$x = 5$

$(5, -2)$

<u>To Think About</u>

25. $cx + dy = -3$

$dx - cy = -3$

$x = 1, \ y = 2$

$c + 2d = -3$ (1)

$-2c + d = -3$ (2)

Multiply (1) by 2

$2c + 4d = -6$

$\underline{-2c + d = -3}$

$5d = -9$

$d = -\dfrac{9}{5}$

25. continued

Substitute $-\dfrac{9}{5}$ for d in (1)

$$c + 2\left(-\dfrac{9}{5}\right) = -3$$

$$c = \dfrac{3}{5}$$

$$d = -\dfrac{9}{5}, \ c = \dfrac{3}{5}$$

27. $\ 4(x+3) + 3(x-4) = -3 + 2(x-1)$

$$4x + 12 + 3x - 12 = -3 + 2x - 2$$

$$7x = 2x - 5$$

$$5x = -5$$

$$x = -1$$

29. $\ 20x + 12\left(\dfrac{9}{2} - x\right) = 70$

$$20x + 54 - 12x = 70$$

$$8x = 16$$

$$x = 2$$

Exercises 8.5

1. $x =$ larger number

   $y =$ smaller number

   $x + y = -4$ (1)

   $\underline{x - y = 30}$ (2)

   $2x \quad = 26$

   $\quad\ x = 13.$

   Substitute 13 for x in (1)

   $13 + y = -4$

   $\quad\ \ y = -17$

   Numbers are 13, -17

3. $x =$ highway mileage

   $y =$ city mileage

   $x = y + 12$ (1)

   $\dfrac{x + y}{2} = 36$ (2)

   Substitute $y + 12$ for x in (2)

   $\dfrac{y + 12 + y}{2} = 36$

   $2y + 12 = 72$

   $2y = 60$

   $y = 30$

   Substitute 30 for y in (1)

   $x = 30 + 12$

   $x = 42$

   City mileage $= 30$ mpg

   Highway mileage $= 42$ mpg

5. $x =$ managers that attended college

   $y =$ managers that did not attend

   $x + y = 46$ (1)

   $\underline{x - y = 18}$ (2)

   $2x \quad = 64$

   $\quad x = 32$

   Substitute 32 for x in (1)

   $32 + y = 46$

   $\quad\ \ y = 14$

   32 attended college

   14 did not attend college

7. $x =$ number of orchestra tickets

   $y =$ number of mezzanine

   $x + y = 420$ (1)

   $13x + 8y = 4610$ (2)

   Multiply (1) by -13

   $-13x - 13y = -5460$

   $\underline{13x + \ \ 8y = 4610}$

   $\quad\quad -5y = -850$

   $\quad\quad\quad y = 170$

7. continued

Substitute 170 for y in (1)

$x + 170 = 420$

$\quad x = 250$

250 orchestra tickets

170 mezzanine tickets

9. w = original width

l = original length

$2w + 2l = 48 \qquad (1)$

$6w + 4l = 118 \qquad (2)$

Multiply (1) by -2

$-4w - 4l = -96$

$\underline{6w + 4l = 118}$

$2w \qquad = 22$

$\qquad w = 11$

Substitute 11 for w in (1)

$2(11) + 2l = 48$

$\qquad 2l = 26$

$\qquad l = 13$

original width = 11 feet

original length = 13 feet

11. x = quarts before

y = quarts added

$x + \quad y = 16 \qquad (1)$

$0.50x + 0.80y = 0.65(16) \qquad (2)$

Multiply (1) by -5 and (2) by 10

$-5x - 5y = -80$

$\underline{5x + 8y = 104}$

$3y = 24$

$y = 8$

Substitute 8 for y in (1)

$x + 8 = 16$

$\quad x = 8$

8 quarts before

8 quarts added

13. x = weight of nuts

y = weight of raisins

$x + \quad y = 50 \qquad (1)$

$2.00x + 1.50y = 50(1.80) \qquad (2)$

Multiply (1) by -2

$-2x - 2y = -100$

$\underline{2x + 1.5y = 90}$

$-0.5y = -10$

$\quad y = 20$

Substitute 20 for y in (1)

$x + 20 = 50$

$\quad x = 30$

30 lbs of nuts

20 lbs of raisins

15. w = speed of wind

r = airspeed of airplane

$\text{speed} = \dfrac{\text{distance}}{\text{time}}$

$r + w = \dfrac{630}{3} \qquad (1)$

$r - w = \dfrac{630}{3.5} \qquad (2)$

$\overline{}$

$2r = 390$

$r = 195$

Substitute 195 for r in (1)

$195 + w = 210$

$\quad w = 15$

Airspeed = 195 mph

Windspeed = 15 mph

17. c = speed of current

r = speed of still water

$r + c = \dfrac{45}{3} \qquad (1)$

$r - c = \dfrac{45}{4} \qquad (2)$

$\overline{}$

$2r \quad = 26.25$

$\quad r = 13.125$

17. continued

Substitute 13.125 for r in (1)

$13.125 + c = 15$

$\qquad c = 1.875$

Speed in still water $= 13.125$ mph

Speed of current $= 1.875$ mph

19. $x =$ value of dictionaries

$y =$ value of encyclopedias

$0.06x + 0.08y = 350 \qquad (1)$

$0.08x + 0.10y = 440 \qquad (2)$

Multiply (1) by 400 and (2) by -300

$24x + 32y = 140,000$

$\underline{-24x - 30y = -132,000}$

$\qquad 2y = 8000$

$\qquad y = 4000$

Substitute 4000 for y in (2)

$0.08x + 0.10(4000) = 440$

$\qquad 0.08x = 40$

$\qquad x = 500$

$500 worth of dictionaries

$4000 worth of encyclopedias

Cumulative Review Problems

21. $2x^2 - 3y + 4y^2, \quad x = -2, \quad y = 3$

$2(-2)^2 - 3(3) + 4(3)^2 = 35$

Putting Your Skills To Work

1. $D =$ number of double rooms

$S =$ number of single rooms

$60D + 52S = 9480 \qquad (1)$

$64D + 49S = 9530 \qquad (2)$

Multiply (1) by 16 and (2) by -15

$960D + 832S = 151,680$

$\underline{-960D - 735S = -142,950}$

$\qquad 97S = 8730$

$\qquad S = 90$

Substitute 90 for S in (1)

$60D + 52(90) = 9480$

$\qquad 60D = 4800$

$\qquad D = 80$

80 double rooms

90 single rooms

3. $70(0.90)D + 40(1.50)S$

$= 63D + 60S$

They would receive $3 more per double and $8 more per single. Instead of receipts of $9480 on August 1, the hotel would obtain receipts of $10,440.

Chapter 8 - Review Problems

1. $3x + y = 4$

$\qquad 3x - 2y = 10$

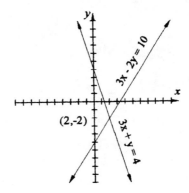

153

3. $-4x + 3y = 15$
   $2x + y = -5$

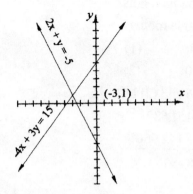

$2x + y = -5$
$(-3, 1)$
$4x + 3y = 15$

5. $x + 3y = 18$ (1)
   $2x + y = 11$ (2)
   Solve (2) for y
   $y = -2x + 11$ (3)
   Substitute $-2x + 11$ for y in (1)
   $x + 3(-2x + 11) = 18$
   $x - 6x + 33 = 18$
   $-5x = -15$
   $x = 3$
   Substitute 3 for x in (3)
   $y = -2(3) + 11 = 5$
   $(3, 5)$

   Check: $3 + 3(5) \overset{?}{=} 18$
   $18 = 18$

7. $3x + 2y = 0$ (1)
   $2x - y = 7$ (2)
   Solve (2) for y
   $y = 2x - 7$ (3)
   Substitute $2x - 7$ for y in (1)
   $3x + 2(2x - 7) = 0$
   $3x + 4x - 14 = 0$
   $7x = 14$
   $x = 2$
   Substitute 2 for x in (3)
   $y = 2(2) - 7 = -3$
   $(2, -3)$

   Check: $3(2) + 2(-3) \overset{?}{=} 0$
   $0 = 0$

9. $6x - 2y = 10$ (1)
   $2x + 3y = 7$ (2)
   Multiply (1) by 3 and (2) by 2
   $18x - 6y = 30$
   $\underline{4x + 6y = 14}$
   $22x \quad\quad = 44$
   $x = 2$
   Substitute 2 for x in (2)
   $2(2) + 3y = 7$
   $3y = 3$
   $y = 1$
   $(2, 1)$

   Check: $6(2) - 2(1) \overset{?}{=} 10$
   $10 = 10$

11. $-5x + 8y = 4$ (1)

    $2x - 7y = 6$ (2)

    Multiply (1) by 2 and (2) by 5

$$-10x + 16y = 8$$
$$\underline{10x - 35y = 30}$$
$$-19y = 38$$
$$y = -2$$

    Substitute -2 for y in (2)

$$2x - 7(-2) = 6$$
$$2x = -8$$
$$x = -4$$
$$(-4, -2)$$

    Check: $\ -5(-4) + 8(-2) \overset{?}{=} 4$

$$4 = 4$$

13. $3x - 2y = 17$ (1)

    $x + \ y = 9$ (2)

    Multiply (2) by 2

$$3x - 2y = 17$$
$$\underline{2x + 2y = 18}$$
$$5x \quad = 35$$
$$x = 7$$

    Substitute 7 for x in (2)

$$7 + y = 9$$
$$y = 2$$
$$(7, 2)$$

15. $11x + 3y = 12$ (1)

    $4x + y = 5$ (2)

    Multiply (2) by -3

$$11x + 3y = 12$$
$$\underline{-12x - 3y = -15}$$
$$-x \quad = -3$$
$$x = 3$$

15. continued

    Substitute 3 for x in (2)

$$4(3) + y = 5$$
$$y = -7$$
$$(3, -7)$$

17. $5x - 3y = 32$ (1)

    $5x + 7y = -8$ (2)

    Multiply (2) by -1

$$5x - 3y = 32$$
$$\underline{-5x - 7y = 8}$$
$$-10y = 40$$
$$y = -4$$

    Substitute -4 for y in (1)

$$5x - 3(-4) = 32$$
$$5x = 20$$
$$x = 4$$
$$(4, -4)$$

19. $7x + 3y = 2$ (1)

    $-8x - 7y = 2$ (2)

    Multiply (1) by 7 and (2) by 3

$$49x + 21y = 14$$
$$\underline{-24x - 21y = 6}$$
$$25x \quad = 20$$
$$x = \frac{4}{5}$$

    Substitute $\frac{4}{5}$ for x in (1)

$$7\left(\frac{4}{5}\right) + 3y = 2$$
$$3y = -\frac{18}{5}$$
$$y = -\frac{6}{5}$$
$$\left(\frac{4}{5}, -\frac{6}{5}\right)$$

21.

$5x - 11y = -4 \qquad (1)$

$6x - 8y = -10 \qquad (2)$

Multiply (1) by 6 and (2) by -5

$30x - 66y = -24$

$\underline{-30x + 40y = 50}$

$\qquad -26y = 26$

$\qquad y = -1$

Substitute -1 for y in (1)

$5x - 11(-1) = -4$

$\qquad 5x = -15$

$\qquad x = -3$

$(-3, -1)$

23.

$2x - 4y = 6 \qquad (1)$

$-3x + 6y = 7 \qquad (2)$

Multiply (1) by 3 and (2) by 2

$6x - 12y = 18$

$\underline{-6x + 12y = 14}$

$\qquad 0 = 32$

No solution, inconsistent system.

25.

$2x - 9y = 0 \qquad (1)$

$3x + 5 = 6y \qquad (2)$

Rearrange (2)

$3x - 6y = -5 \qquad (3)$

Multiply (1) by 3 and (3) by -2

$6x - 27y = 0$

$\underline{-6x + 12y = 10}$

$\qquad -15y = 10$

$\qquad y = -\dfrac{2}{3}$

Substitute $-\dfrac{2}{3}$ for y in (1)

$2x - 9\left(-\dfrac{2}{3}\right) = 0$

$\qquad 2x = -6$

$\qquad x = -3$

$\left(-3, -\dfrac{2}{3}\right)$

27.

$5x - 8y = 3x + 12 \qquad (1)$

$7x + y = 6y - 4 \qquad (2)$

Rearrange and simplify

$2x - 8y = 12 \qquad (3)$

$7x - 5y = -4 \qquad (4)$

Multiply (3) by 7 and (4) by -2

$14x - 56y = 84$

$\underline{-14x + 10y = 8}$

$\qquad -46y = 92$

$\qquad y = -2$

Substitute -2 for y in (3)

$2x - 8(-2) = 12$

$\qquad 2x = -4$

$\qquad x = -2$

$(-2, -2)$

29.

$3x + y = 9 \qquad (1)$

$x - 2y = 10 \qquad (2)$

Multiply (1) by 2

$6x + 2y = 18$

$\underline{x - 2y = 10}$

$7x \qquad = 28$

$\qquad x = 4$

Substitute 4 for x in (1)

$3(4) + y = 9$

$\qquad y = -3$

$(4, -3)$

31.

$2(x + 3) = y + 4 \qquad (1)$

$4x - 2y = -4 \qquad (2)$

Expand, rearrange and simplify

$2x - y = -2 \qquad (3)$

$4x - 2y = -4 \qquad (4)$

31. continued

Multiply (3) by -2

$-4x + 2y = 4$

$\underline{4x - 2y = -4}$

$\quad\quad 0 = 0$

Infinite number of solutions, dependent system.

33. $4x - 3y + 1 = 6$ (1)

$5x + 8y + 2 = -74$ (2)

Rearrange and simplify

$4x - 3y = 5$ (3)

$5x + 8y = -76$ (4)

Multiply (3) by 8 and (4) by 3

$32x - 24y = 40$

$\underline{15x + 24y = -228}$

$\quad\quad 47x = -188$

$\quad\quad\quad x = -4$

Substitute -4 for x in (2)

$5(-4) + 8y + 2 = -74$

$\quad\quad\quad 8y = -56$

$\quad\quad\quad\quad y = -7$

$(-4, -7)$

35. $\dfrac{2x}{3} - \dfrac{3y}{4} = \dfrac{7}{12}$ (1)

$8x + 5y = 9$ (2)

Multiply (1) by -12

$-8x + 9y = -7$

$\underline{8x + 5y = 9}$

$\quad\quad 14y = 2$

$\quad\quad\quad y = \dfrac{1}{7}$

35. continued

Substitute $\dfrac{1}{7}$ for y in (2)

$8x + 5\left(\dfrac{1}{7}\right) = 9$

$\quad\quad 8x = \dfrac{58}{7}$

$\quad\quad\quad x = \dfrac{29}{28}$

$\left(\dfrac{29}{28}, \dfrac{1}{7}\right)$

37. $\dfrac{1}{5}a + \dfrac{1}{2}b = 6$ (1)

$\dfrac{3}{5}a - \dfrac{1}{2}b = 2$ (2)

$\dfrac{4}{5}a \quad\quad = 8$

$\quad\quad a = 10$

Substitute 10 for a in (1)

$\dfrac{1}{5}(10) + \dfrac{1}{2}b = 6$

$\quad\quad \dfrac{1}{2}b = 4$

$\quad\quad\quad b = 8$

$(10, 8)$

39. $0.2s - 0.3t = 0.3$ (1)

$0.4s + 0.6t = -0.2$ (2)

Multiply (1) by 20 and (2) by 10

$4s - 6t = 6$ (3)

$\underline{4s + 6t = -2}$ (4)

$8s \quad\quad = 4$

$\quad s = \dfrac{1}{2}$

39. continued

Substitute $\frac{1}{2}$ for s in (4)

$$4\left(\frac{1}{2}\right) + 6t = -2$$

$$6t = -4$$

$$t = -\frac{2}{3}$$

$$\left(\frac{1}{2}, -\frac{2}{3}\right)$$

41.
$$3m + 2n = 5 \qquad (1)$$
$$4n = 10 - 6m \qquad (2)$$

Rearrange

$$3m + 2n = 5 \qquad (3)$$
$$6m + 4n = 10 \qquad (4)$$

Multiply (3) by -2

$$\begin{array}{l} -6m - 4n = -10 \\ \underline{6m + 4n = 10} \\ 0 = 0 \end{array}$$

Infinite number of solutions,
dependent system.

43.
$$3(x + 2) = -2 - (x + 3y) \qquad (1)$$
$$3(x + y) = 3 - 2(y - 1) \qquad (2)$$

Expand, rearrange and simplify

$$4x + 3y = -8 \qquad (3)$$
$$3x + 5y = 5 \qquad (4)$$

Multiply (3) by 3 and (4) by -4

$$\begin{array}{l} 12x + 9y = -24 \\ \underline{-12x - 20y = -20} \\ -11y = -44 \\ y = 4 \end{array}$$

43. continued

Substitute 4 for y in (3)

$$4x + 3(4) = -8$$
$$4x = -20$$
$$x = -5$$

$$(-5, 4)$$

45.
$$0.2b = 1.4 - 0.3a \qquad (1)$$
$$0.1b + 0.6 = 0.5a \qquad (2)$$

Rearrange

$$0.3a + 0.2b = 1.4 \qquad (3)$$
$$0.5a - 0.1b = 0.6 \qquad (4)$$

Multiply (3) by 10 and (4) by 20

$$\begin{array}{l} 3a + 2b = 14 \\ \underline{10a - 2b = 12} \\ 13a = 26 \\ a = 2 \end{array}$$

Substitute 2 for a in (1)

$$0.2b = 1.4 - 0.3(2)$$
$$b = 4$$

$$(2, 4)$$

47.
$$\frac{b}{5} = \frac{2}{5} - \frac{a-3}{2} \qquad (1)$$
$$4(a - b) = 3b - 2(a - 2) \qquad (2)$$

Multiply (1) by 10, then
Expand, rearrange, and simplify.

$$5a + 2b = 19 \qquad (3)$$
$$6a - 7b = 4 \qquad (4)$$

Multiply (3) by 7 and (4) by 2

$$\begin{array}{l} 35a + 14b = 133 \\ \underline{12a - 14b = 8} \\ 47a = 141 \\ a = 3 \end{array}$$

47. continued

Substitute 3 for a in (3)

$5(3) + 2b = 19$

$2b = 4$

$b = 2$

$(3, 2)$

49.

$$ax - by = ab + b^2 \qquad (1)$$

$$\underline{ax + by = 2a^2 + ab - b^2 \qquad (2)}$$

$$2ax \quad = 2a^2 + 2ab$$

$$x = a + b$$

Substitute $a + b$ for x in (2)

$$a(a + b) + by = 2a^2 + ab - b^2$$

$$a^2 + ab + by = 2a^2 + ab - b^2$$

$$by = a^2 - b^2$$

$$y = \frac{a^2 - b^2}{b}$$

$$\left(a + b, \ \frac{a^2 - b^2}{b} \right)$$

51. x = number of reserved seat tickets

y = number of general admission tickets

$$x + y = 760 \qquad (1)$$

$$6x + 4y = 3280 \qquad (2)$$

Multiply (1) by -4

$$-4x - 4y = -3040$$

$$\underline{6x + 4y = 3280}$$

$$2x \qquad = 240$$

$$x = 120$$

Substitute 120 for x in (1)

$$120 + y = 760$$

$$y = 640$$

120 reserved seats

640 general admission

53. r = airspeed

w = wind speed

$$\text{speed} = \frac{\text{distance}}{\text{time}}$$

$$r + w = \frac{1500}{5} \qquad (1)$$

$$\underline{r - w = \frac{1500}{6} \qquad (2)}$$

$$2r \quad = 550$$

$$r = 275$$

Substitute 275 for r in (1)

$$275 + w = 300$$

$$w = 25$$

Speed in still air = 275 mph

Wind speed = 25 mph

55. x = number of regular - sized cars

y = number of compact cars

$$x + y = 86 \qquad (1)$$

$$4x + 3.50y = 323 \qquad (2)$$

Multiply (1) by -4

$$-4x - 4y = -344$$

$$\underline{4x + 3.50y = 323}$$

$$-0.50y = -21$$

$$y = 42$$

Substitute 42 for y in (1)

$$x + 42 = 86$$

$$x = 44$$

44 regular - sized cars

42 compact cars

57. r = speed of boat in still water

c = speed of current

$r + c = 23$ (1)

$\underline{r - c = 15}$ (2)

$2r\ \ = 38$

$\ \ \ r = 19$

Substitute 19 for r in (1)

$19 + c = 23$

$\ \ \ \ \ c = 4$

Speed of boat = 19 kph

Speed of current = 4 kph

Putting Your Skills To Work

1. $C = 300 + 2x,\ \ R = 3x,\ \ x = 200$

(a) $C = 300 + 2(200) = \$700$

(b) $R = 3(200) = \$600$

(c) This is a loss of \$100 because it cost \$100 more than you make.

3. $C = 300 + 2x,\ \ R = 3x,\ \ x = 275$

$C = 300 + 2(275) = \$850$

$R = 3(275) = \$825$

Profit $= 850 - 825 = \$25$

5.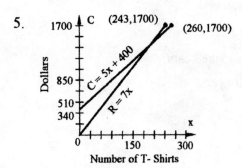

7. $x = 320$

$C = 400 + 5x = 400 + 5(320) = 2000$

$R = 7x = 7(320) = 2240$

$R - C = 2240 - 2000 = \$240$ Profit

Test - Chapter 8

1. $3x -\ \ y = -5$ (1)

$-2x + 5y = -14$ (2)

Solve (1) for y

$y = 3x + 5$ (3)

Substitute $3x + 5$ for y in (2)

$-2x + 5(3x + 5) = -14$

$-2x + 15x + 25 = -14$

$\ \ \ \ \ \ \ \ \ \ \ 13x = -39$

$\ \ \ \ \ \ \ \ \ \ \ \ \ \ x = -3$

Substitute -3 for x in (1)

$3(-3) - y = -5$

$\ \ \ \ \ \ \ -y = 4$

$\ \ \ \ \ \ \ \ \ y = -4$

$(-3, -4)$

3. $2x - y = 4$

$4x + y = 2$

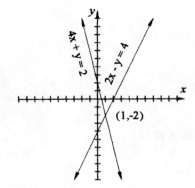

5. $2x - y = 5$ (1)

   $-x + 3y = 5$ (2)

   Multiply (1) by 3

   $6x - 3y = 15$

   $\underline{-x + 3y = 5}$

   $5x \qquad = 20$

         $x = 4$

   Substitute 4 for x in (1)

   $2(4) - y = 5$

       $-y = -3$

         $y = 3$

   $(4, 3)$

7. $\dfrac{2}{3}x - \dfrac{1}{5}y = 2$ (1)

   $\dfrac{4}{3}x + 4y = 4$ (2)

   Multiply (1) by 15 and (2) by 3

   $10x - 3y = 30$ (3)

   $4x + 12y = 12$ (4)

   Multiply (3) by 4

   $40x - 12y = 120$

   $\underline{4x + 12y = 12}$

   $44x \qquad = 132$

        $x = 3$

   Substitute 3 for x in (2)

   $\dfrac{4}{3}(3) + 4y = 4$

         $4y = 0$

          $y = 0$

9. $5x - 2 = y$ (1)

   $10x = 4 + 2y$ (2)

   Substitute $5x - 2$ for y in (2)

   $10x = 4 + 2(5x - 2)$

   $10x = 4 + 10x - 4$

      $0 = 0$

   Infinite number of solutions,

   dependent system.

11. $2(x + y) = 2(1 - y)$ (1)

   $5(-x + y) = 2(23 - y)$ (2)

   Expand, rearrange, and simplify

   $2x + 4y = 2$ (3)

   $-5x + 7y = 46$ (4)

   Multiply (3) by 5 and (4) by 2

   $10x + 20y = 10$

   $\underline{-10x + 14y = 92}$

        $34y = 102$

          $y = 3$

   Substitute 3 for y in (3)

   $2x + 4(3) = 2$

        $2x = -10$

         $x = -5$

   $(-5, 3)$

13. $x =$ first number

   $y =$ second number

   $2x + 3y = 1$ (1)

   $3x + 2y = 9$ (2)

   Multiply (1) by 3 and (2) by -2

   $6x + 9y = 3$

   $\underline{-6x - 4y = -18}$

        $5y = -15$

         $y = -3$

   Substitute -3 for y in (1)

   $2x + 3(-3) = 1$

        $2x = 10$

         $x = 5$

   $(5, -3)$

15. x = cost of a shirt

y = cost of a pair of slacks

$5x + 3y = 172$ (1)

$3x + 4y = 156$ (2)

Multiply (1) by 4 and (2) by -3

$20x + 12y = 688$

$\underline{-9x - 12y = -468}$

$11x \qquad = 220$

$\qquad x = 20$

Substitute 20 for x in (1)

$5(20) + 3y = 172$

$\qquad 3y = 72$

$\qquad y = 24$

A shirt cost $20

A pair of slacks costs $24

17. r = airspeed

w = wind speed

$r + w = \dfrac{2000}{4}$ (1)

$r - w = \dfrac{2000}{5}$ (2)

$\overline{2r \quad = 900}$

$\quad r = 450$

Substitute 450 for r in (1)

$450 + w = 500$

$\qquad w = 50$

Speed of plane in still air = 450 kmph

Wind speed = 50 kmph

Cumulative Test - Chapters 0 - 8

1. 11% of $37.20 = 0.11(37.20) = \$4.09$

3. $(3x^2 - 2x + 1)(5x - 3)$

$= (3x^2 - 2x + 1)(5x) - (3x^2 - 2x + 1)(3)$

$= 15x^3 - 10x^2 + 5x - 9x^2 + 6x - 3$

$= 15x^3 - 19x^2 + 11x - 3$

5. $5(3 - x) \geq 6$

$15 - 5x \geq 6$

$\quad -5x \geq -9$

$\qquad x \leq \dfrac{9}{5}$

7. 12% of x is 360

$0.12x = 360$

$\quad x = 3000$

9. $5a^3 - 5a^2 - 60a$

$= 5a(a^2 - a - 12)$

$= 5a(a - 4)(a + 3)$

11. $2x + \ y = 8$ (1)

$3x + 4y = -8$ (2)

Multiply (1) by -4

$-8x - 4y = -32$

$\underline{3x + 4y = -8}$

$-5x \qquad = -40$

$\qquad x = 8$

Substitute 8 for x in (1)

$2(8) + y = 8$

$\qquad y = -8$

$(8, -8)$

13. $1.3x - 0.7y = 0.4$ (1)

$-3.9x + 2.1y = -1.2$ (2)

Multiply (1) by 30 and (2) by 10

$39x - 21y = 12$

$\underline{-39x + 21y = -12}$

$\qquad 0 = 0$

Infinite number of solutions, dependent system.

15. Parallel lines represent an inconsistent system for which there is no solution.

17. r = speed of boat in still water

s = speed of the stream

$$r - s = \frac{6}{3} \quad (1)$$

$$r + s = \frac{6}{1.5} \quad (2)$$

Multiply (1) by -1

$-r + s = -2$

$\underline{r + s = 4}$

$2s = 2$

$s = 1$

Speed of the stream is 1 mph

Practice Quiz: Sections 8.1 - 8.2

1. Solve by graphing.

$3x - 4y = 5$

$-9x + 12y = 3$

2. Solve by graphing.

$2x + 3y = 1$

$5x + 3y = 16$

3. Solve by the substitution method.

$x - 2y = 8$

$3x - 2y = 12$

4. Solve by the substitution method.

$2x + y = 8$

$-4x - 2y = -16$

Practice Quiz: Sections 8.3 - 8.5

1. Solve by the addition method.

$5x - 2y = 31$

$4x + 3y = 11$

2. Solve by the addition method.

$-2x + 7y = 2$

$3x - 5y = -14$

3. Solve by the addition method

$3x - 2y = 7$

$-6x + 4y = -15$

4. A purchase of 6 packages of paper plates and 4 packages of paper cups cost \$17, while a purchase of 4 packages of plates and 6 packages of cups cost \$18. What is the cost per package of plates and the cost per package of cups?

Answers to Practice Quiz

Sections 8.1 - 8.2

1.
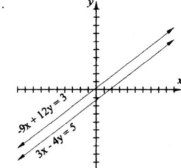

No solution, inconsistent system.

2.
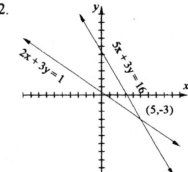

3. (2,-3)

4. Infinite number of solutions, dependent system.

163

Answers to Practice Quiz

Sections 8.3 - 8.5

1. $(5, -3)$

2. $(-8, -2)$

3. No solution, inconsistent system

4. Plates are $1.50 and cups are $2.00.

1. $\sqrt{121} = 11$

3. $\sqrt{169} = 13$

5. $\sqrt{5} = 2.236$

7. $\sqrt{x^4} = x^2$

9. $\sqrt{98} = \sqrt{49 \cdot 2} = 7\sqrt{2}$

11. $\sqrt{36x^3} = \sqrt{36x^2 x} = 6x\sqrt{x}$

13. $5\sqrt{2} - \sqrt{2} = 4\sqrt{2}$

15. $3\sqrt{2} - \sqrt{8} + \sqrt{18} = 3\sqrt{2} - \sqrt{4 \cdot 2} + \sqrt{9 \cdot 2}$
$$= 3\sqrt{2} - 2\sqrt{2} + 3\sqrt{2}$$
$$= 4\sqrt{2}$$

17. $3\sqrt{8} + \sqrt{12} + \sqrt{50} - 4\sqrt{75}$
$$= 3\sqrt{4 \cdot 2} + \sqrt{4 \cdot 3} + \sqrt{25 \cdot 2} - 4\sqrt{25 \cdot 3}$$
$$= 6\sqrt{2} + 2\sqrt{3} + 5\sqrt{2} - 20\sqrt{3}$$
$$= 11\sqrt{2} - 18\sqrt{3}$$

19. $\sqrt{2}\left(\sqrt{8} + 2\sqrt{3}\right) = \sqrt{16} + 2\sqrt{6}$
$$= 4 + 2\sqrt{6}$$

21. $\left(\sqrt{11} - \sqrt{10}\right)\left(\sqrt{11} + \sqrt{10}\right)$
$$= \sqrt{121} - \sqrt{100}$$
$$= 11 - 10$$
$$= 1$$

23. $\left(2\sqrt{3} - \sqrt{5}\right)^2 = 4\sqrt{9} + 2\left(-\sqrt{5}\right)\left(2\sqrt{3}\right) + \sqrt{25}$
$$= 12 - 4\sqrt{15} + 5$$
$$= 17 - 4\sqrt{15}$$

25. $\dfrac{5}{\sqrt{3}} = \dfrac{5}{\sqrt{3}} \cdot \dfrac{\sqrt{3}}{\sqrt{3}} = \dfrac{5\sqrt{3}}{3}$

27. $\dfrac{\sqrt{7} - \sqrt{6}}{\sqrt{7} + \sqrt{6}} = \dfrac{\sqrt{7} - \sqrt{6}}{\sqrt{7} + \sqrt{6}} \cdot \dfrac{\sqrt{7} - \sqrt{6}}{\sqrt{7} - \sqrt{6}}$
$$= \dfrac{\sqrt{49} + 2\left(-\sqrt{6}\right)\left(\sqrt{7}\right) + \sqrt{36}}{\left(\sqrt{7}\right)^2 - \left(\sqrt{6}\right)^2}$$
$$= \dfrac{7 - 2\sqrt{42} + 6}{1}$$
$$= 13 - 2\sqrt{42}$$

29. $6^2 = x^2 + 5^2$
$$36 = x^2 + 25$$
$$11 = x^2$$
$$\sqrt{11} = x$$

31. $\sqrt{3x + 1} = 6$
$$3x + 1 = 36$$
$$3x = 35$$
$$x = \dfrac{35}{3}$$

33. $y = \dfrac{k}{x}, \quad y = \dfrac{2}{3}, \quad x = 21$
$$\dfrac{2}{3} = \dfrac{k}{21} \Rightarrow k = 14$$
$$y = \dfrac{14}{x}, \quad x = 2$$
$$y = \dfrac{14}{2} = 7$$

Exercises 9.1

1. Answers will vary.

3. Yes, $\sqrt{900} = 30$ because $(30)(30) = 900$

5. $\pm\sqrt{9} = \pm 3$

7. $\pm\sqrt{64} = \pm 8$

9. $\pm\sqrt{81} = \pm 9$

11. $\pm\sqrt{121} = \pm11$

13. $\sqrt{25} = 5$

15. $\sqrt{0} = 0$

17. $\sqrt{49} = 7$

19. $\sqrt{100} = 10$

21. $-\sqrt{36} = -6$

23. $\sqrt{0.81} = 0.9$

25. $\sqrt{0.01} = 0.1$

27. $\sqrt{\dfrac{16}{25}} = \dfrac{4}{5}$

29. $\sqrt{\dfrac{49}{64}} = \dfrac{7}{8}$

31. $\sqrt{400} = 20$

33. $-\sqrt{10,000} = -100$

35. $\sqrt{169} = 13$

37. $-\sqrt{\dfrac{1}{64}} = -\dfrac{1}{8}$

39. $\sqrt{\dfrac{25}{49}} = \dfrac{5}{7}$

41. $\sqrt{0.36} = 0.6$

43. $\sqrt{14,400} = 120$

45. $\sqrt{441} = 21$

47. $\sqrt{0.0064} = 0.08$

49. $\sqrt{42} = 6.481$

51. $\sqrt{59} = 7.681$

53. $-\sqrt{133} = -11.533$

55. $-\sqrt{195} = -13.964$

57. $-\sqrt{360} = -18.97$

59. $\sqrt{45.5} = 6.75$

61. If $a = 900$,
$$t = \frac{1}{4}\sqrt{a} = \frac{1}{4}\sqrt{900} = 7.5 \text{ seconds}$$

63. If $a = 1444$,
$$t = \frac{1}{4}\sqrt{a} = \frac{1}{4}\sqrt{1444} = 9.5 \text{ seconds}$$

To Think About

65. $\sqrt{\sqrt{81}} = \sqrt{9} = 3$

Cumulative Review Problems

67. $3x + 2y = 8 \qquad (1)$
$7x - 3y = 11 \qquad (2)$
Multiply (1) by 3 and (2) by 2
$9x + 6y = 24$
$\underline{14x - 6y = 22}$
$23x \qquad = 46$
$x = 2$
Substitute 2 for x in (1)
$3(2) + 2y = 8$
$2y = 2$
$y = 1$
$(2, 1)$

166

69. t = cost per tire

b = cost per battery

$3t + 2b = 360$ (1)

$4t + 3b = 500$ (2)

Multiply (1) by 3 and (2) by -2

$9t + 6b = 1080$

$\underline{-8t - 6b = -1000}$

$t \quad\quad = 80$

Substitute 80 for t in (1)

$3(80) + 2b = 360$

$2b = 120$

$b = 60$

$80 per tire and $60 per battery

Exercises 9.2

1. $\sqrt{7^2} = 7$

3. $\sqrt{10^4} = \sqrt{\left(10^2\right)^2} = 10^2$

5. $\sqrt{5^{10}} = \sqrt{\left(5^5\right)^2} = 5^5$

7. $\sqrt{56^6} = \sqrt{\left(56^3\right)^2} = 56^3$

9. $\sqrt{3^{150}} = \sqrt{\left(3^{75}\right)^2} = 3^{75}$

11. $\sqrt{w^{12}} = \sqrt{\left(w^6\right)^2} = w^6$

13. $\sqrt{x^{24}} = \sqrt{\left(x^{12}\right)^2} = x^{12}$

15. $\sqrt{y^{26}} = \sqrt{\left(y^{13}\right)^2} = y^{13}$

17. $\sqrt{49x^6} = \sqrt{49}\sqrt{x^6} = 7x^3$

19. $\sqrt{144x^2} = \sqrt{144}\sqrt{x^2} = 12x$

21. $\sqrt{400x^8} = \sqrt{400}\sqrt{x^8} = 20x^4$

23. $\sqrt{x^2y^6} = \sqrt{x^2}\sqrt{y^6} = xy^3$

25. $\sqrt{25x^8y^4} = \sqrt{25}\sqrt{x^8}\sqrt{y^4} = 5x^4y^2$

27. $\sqrt{64x^2y^4} = \sqrt{64}\sqrt{x^2}\sqrt{y^4} = 8xy^2$

29. $\sqrt{121x^6y^{14}} = \sqrt{121}\sqrt{x^6}\sqrt{y^{14}} = 11x^3y^7$

31. $\sqrt{20} = \sqrt{4\cdot 5} = 2\sqrt{5}$

33. $\sqrt{24} = \sqrt{4\cdot 6} = 2\sqrt{6}$

35. $\sqrt{18} = \sqrt{9\cdot 2} = 3\sqrt{2}$

37. $\sqrt{50} = \sqrt{25\cdot 2} = 5\sqrt{2}$

39. $\sqrt{72} = \sqrt{36\cdot 2} = 6\sqrt{2}$

41. $\sqrt{90} = \sqrt{9\cdot 10} = 3\sqrt{10}$

43. $\sqrt{98} = \sqrt{49\cdot 2} = 7\sqrt{2}$

45. $\sqrt{150} = \sqrt{25\cdot 6} = 5\sqrt{6}$

47. $\sqrt{12y^5} = \sqrt{4\cdot 3y^4y} = 2y^2\sqrt{3y}$

49. $\sqrt{18y^{13}} = \sqrt{9\cdot 2y^{12}y} = 3y^6\sqrt{2y}$

51. $\sqrt{49x^5} = \sqrt{49x^4x} = 7x^2\sqrt{x}$

53. $\sqrt{50x^7y} = \sqrt{25\cdot 2x^6xy} = 5x^3\sqrt{2xy}$

55. $\sqrt{12x^2y^3} = \sqrt{4\cdot 3x^2y^2y} = 2xy\sqrt{3y}$

57. $\sqrt{27a^8y^7} = \sqrt{9\cdot 3a^8y^6y} = 3a^4y^3\sqrt{3y}$

59. $\sqrt{140x^7y^{11}} = \sqrt{4\cdot 35x^6xy^{10}y} = 2x^3y^5\sqrt{35xy}$

61. $\sqrt{45ab^3c^4} = \sqrt{9\cdot 5ab^2bc^4} = 3bc^2\sqrt{5ab}$

63. $\sqrt{63a^4b^6c^3} = \sqrt{9\cdot 7a^4b^6c^2c} = 3a^2b^3c\sqrt{7c}$

65. $\sqrt{81x^{12}y^{11}w^5} = \sqrt{81x^{12}y^{10}yw^4w}$
$= 9x^6y^5w^2\sqrt{yw}$

To Think About

67. $\sqrt{x^2+12x+36} = \sqrt{(x+6)^2} = x+6$

69. $\sqrt{9x^2+12x+4} = \sqrt{(3x+2)^2} = 3x+2$

71. $\sqrt{16x^2+16x+4} = \sqrt{4(4x^2+4x+1)}$
$= \sqrt{4(2x+1)^2}$
$= 2(2x+1)$

Cumulative Review Problems

73. $\dfrac{1}{3}(2x-4) = \dfrac{1}{4}x-3$
$4(2x-4) = 3x-36$
$8x-16 = 3x-36$
$5x = -20$
$x = -4$

75. $x =$ a number
$y =$ another number
$x+y = 20$ (1)
$2(x-y) = 11$ (2)
Multiply (1) by 2 and expand (2)
$2x+2y = 40$
$\dfrac{2x-2y = 11}{}$
$4x = 51$
$x = \dfrac{51}{4}$
Substitute $\dfrac{51}{4}$ for x in (1)
$\dfrac{51}{4} + y = 20$
$y = \dfrac{29}{4}$
The numbers are $\dfrac{51}{4}$ and $\dfrac{29}{4}$.

Exercises 9.3

1. (1) Simplify each radical term.
   (2) Combine like radicals.

3. $\sqrt{3} - 5\sqrt{3} + 2\sqrt{3} = -2\sqrt{3}$

5. $\sqrt{2} + 8\sqrt{3} - 5\sqrt{3} + 4\sqrt{2}$
$= 5\sqrt{2} + 3\sqrt{3}$

7. $3\sqrt{2x} + 5\sqrt{x} - 7\sqrt{2x}$
$= 5\sqrt{x} - 4\sqrt{2x}$

9. $\sqrt{3} - \sqrt{12} = \sqrt{3} - \sqrt{4\cdot 3}$
$= 3 - 2\sqrt{3}$
$= -\sqrt{3}$

11. $\sqrt{50} + 3\sqrt{32} = \sqrt{25\cdot 2} + 3\sqrt{16\cdot 2}$
$= 5\sqrt{2} + 12\sqrt{2}$
$= 17\sqrt{2}$

13. $2\sqrt{8} - 3\sqrt{2} = 2\sqrt{4\cdot 2} - 3\sqrt{2}$
$= 4\sqrt{2} - 3\sqrt{2}$
$= \sqrt{2}$

15. $\sqrt{75} + \sqrt{3} - 2\sqrt{27} = \sqrt{25\cdot 3} + \sqrt{3} - 2\sqrt{9\cdot 3}$
$= 5\sqrt{3} + \sqrt{3} - 6\sqrt{3}$
$= 0$

17. $2\sqrt{12} + \sqrt{20} + \sqrt{36} = 2\sqrt{4\cdot 3} + \sqrt{4\cdot 5} + 6$
$= 4\sqrt{3} + 2\sqrt{5} + 6$

19. $\sqrt{20} - \sqrt{80} + 3\sqrt{48} = \sqrt{4\cdot 5} - \sqrt{16\cdot 5} + 3\sqrt{16\cdot 3}$
$= 2\sqrt{5} - 4\sqrt{5} + 12\sqrt{3}$
$= -2\sqrt{5} + 12\sqrt{3}$

21. $3\sqrt{3x} + \sqrt{12x} = 3\sqrt{3x} + \sqrt{4\cdot 3x}$
$= 3\sqrt{3x} + 2\sqrt{3x}$
$= 5\sqrt{3x}$

23. $3\sqrt{8xy} - \sqrt{50xy} = 3\sqrt{4 \cdot 2xy} - \sqrt{25 \cdot 2xy}$
$= 6\sqrt{2xy} - 5\sqrt{2xy}$
$= \sqrt{2xy}$

25. $1.2\sqrt{3x} - 0.5\sqrt{12x} = 1.2\sqrt{3x} - 0.5\sqrt{4 \cdot 3x}$
$= 1.2\sqrt{3x} - \sqrt{3x}$
$= 0.2\sqrt{3x}$

27. $\sqrt{20y} + 2\sqrt{45y} - \sqrt{5y} = \sqrt{4 \cdot 5y} + 2\sqrt{9 \cdot 5y} - \sqrt{5y}$
$= 2\sqrt{5y} + 6\sqrt{5y} - \sqrt{5y}$
$= 7\sqrt{5y}$

29. $3\sqrt{y^3} + 2y\sqrt{y} = 3\sqrt{y^2 y} + 2y\sqrt{y}$
$= 3y\sqrt{y} + 2y\sqrt{y}$
$= 5y\sqrt{y}$

31. $2x\sqrt{8x} + 5\sqrt{18x} = 2x\sqrt{4 \cdot 2x} + 5\sqrt{9 \cdot 2x}$
$= 4x\sqrt{2x} + 15\sqrt{2x}$
$= (4x + 15)\sqrt{2x}$

33. $5\sqrt{8x^3} - 3x\sqrt{50x} = 5\sqrt{4 \cdot 2xx^2} - 3x\sqrt{25 \cdot 2x}$
$= 10x\sqrt{2x} - 15x\sqrt{2x}$
$= -5x\sqrt{2x}$

35. $3\sqrt{27x^2} - 2\sqrt{48x^2} = 3\sqrt{9 \cdot 3x^2} - 2\sqrt{16 \cdot 3x^2}$
$= 9x\sqrt{3} - 8x\sqrt{3}$
$= x\sqrt{3}$

37. $2\sqrt{6y^3} - 2y\sqrt{54} = 2\sqrt{6yy^2} - 2y\sqrt{9 \cdot 6}$
$= 2y\sqrt{6y} - 6y\sqrt{6}$

39. $-5\sqrt{72x^6} - 2\sqrt{50x^6} = -5\sqrt{36 \cdot 2x^6} - 2\sqrt{25 \cdot 2x^6}$
$= -30x^3\sqrt{2} - 10x^3\sqrt{2}$
$= -40x^3\sqrt{2}$

41. $5x\sqrt{8x} - 24\sqrt{50x^3} = 5x\sqrt{4 \cdot 2x} - 24\sqrt{25 \cdot 2xx^2}$
$= 10x\sqrt{2x} - 120x\sqrt{2x}$
$= -110x\sqrt{2x}$

43. 1 short panel: $2(6) + 2(4) + 2\sqrt{52} = 20 + 2\sqrt{52}$
1 long panel: $2(8) + 2(4) + 2\sqrt{80} = 24 + 2\sqrt{80}$

Total yards $= 2(20 + 2\sqrt{52}) + 2(24 + 2\sqrt{80})$
$= 40 + 4\sqrt{4 \cdot 13} + 48 + 4\sqrt{16 \cdot 5}$
$= 88 + 8\sqrt{13} + 16\sqrt{5}$

To Think About

45. $\sqrt{x^5 - 2x^2} + \sqrt{4x^5 - 8x^2}$
$= \sqrt{x^2(x^3 - 2)} + \sqrt{4x^2(x^3 - 2)}$
$= x\sqrt{x^3 - 2} + 2x\sqrt{x^3 - 2}$
$= 3x\sqrt{x^3 - 2}$

47. $\sqrt{x^2 + 2x + 1} - \sqrt{x^2 - 2x + 1}$
$= \sqrt{(x+1)^2} - \sqrt{(x-1)^2}$
$= x + 1 - (x - 1)$
$= 2$

49. $\sqrt{2} + \sqrt{\dfrac{1}{2}} = \sqrt{2} + \sqrt{\dfrac{2}{4}}$
$= \sqrt{2} + \dfrac{\sqrt{2}}{2}$
$= \dfrac{3}{2}\sqrt{2}$

Cumulative Review Problems

51. $3y - 2x \leq 6$

51. continued

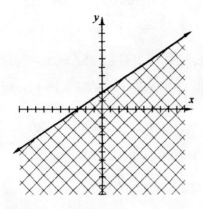

53. $x =$ number of 15 cent stamps

$132 - x =$ number of 25 cent stamps

$0.15x + 0.25(132 - x) = 27.5$

$15x + 0.25(132 - x) = 2750$

$15x + 3300 - 25x = 2750$

$-10x = -550$

$x = 55$, and

$132 - x = 77$

55 of the 15 cent stamps

77 of the 25 cent stamps

Exercises 9.4

1. $\sqrt{7}\sqrt{5} = \sqrt{35}$

3. $\sqrt{2}\sqrt{22} = \sqrt{44} = \sqrt{4 \cdot 11} = 2\sqrt{11}$

5. $\sqrt{3a}\sqrt{5a} = \sqrt{15a^2} = a\sqrt{15}$

7. $\sqrt{5x}\sqrt{10} = \sqrt{50x} = \sqrt{25 \cdot 2x} = 5\sqrt{2x}$

9. $\left(3\sqrt{5}\right)\left(2\sqrt{6}\right) = 6\sqrt{30}$

11. $\left(2x\sqrt{x}\right)\left(3x\sqrt{5x}\right) = 6x^2\sqrt{5x^2}$

$= 6x^3\sqrt{5}$

13. $\left(-3\sqrt{ab}\right)\left(2\sqrt{b}\right) = -6\sqrt{ab^2}$

$= -6b\sqrt{a}$

15. $\sqrt{3}\left(\sqrt{2} + 2\sqrt{5}\right) = \sqrt{6} + 2\sqrt{15}$

17. $\sqrt{2}\left(\sqrt{10} - 3\sqrt{6}\right) = \sqrt{20} - 3\sqrt{12}$

$= \sqrt{4 \cdot 5} - 3\sqrt{4 \cdot 3}$

$= 2\sqrt{5} - 6\sqrt{3}$

19. $2\sqrt{x}\left(\sqrt{x} - 8\sqrt{5}\right) = 2\sqrt{x^2} - 16\sqrt{5x}$

$= 2x - 16\sqrt{5x}$

21. $\sqrt{6}\left(\sqrt{2} - 3\sqrt{6} + 2\sqrt{10}\right) = \sqrt{12} - 3\sqrt{36} + 2\sqrt{60}$

$= \sqrt{4 \cdot 3} - 18 + 2\sqrt{4 \cdot 15}$

$= 2\sqrt{3} - 18 + 4\sqrt{15}$

23. $2\sqrt{a}\left(3\sqrt{b} + \sqrt{ab} - 2\sqrt{a}\right)$

$= 6\sqrt{ab} + 2\sqrt{a^2b} - 4\sqrt{a^2}$

$= 6\sqrt{ab} + 2a\sqrt{b} - 4a$

25. $\left(\sqrt{7} + \sqrt{2}\right)\left(2\sqrt{7} + \sqrt{2}\right)$

$= 2\sqrt{49} + \sqrt{14} + 2\sqrt{14} + \sqrt{4}$

$= 14 + 3\sqrt{14} + 2$

$= 16 + 3\sqrt{14}$

27. $\left(8 + \sqrt{3}\right)\left(2 + 2\sqrt{3}\right)$

$= 16 + 16\sqrt{3} + 2\sqrt{3} + 2\sqrt{9}$

$= 16 + 18\sqrt{3} + 6$

$= 22 + 18\sqrt{3}$

29. $\left(2\sqrt{7} - 3\sqrt{3}\right)\left(\sqrt{7} + \sqrt{3}\right)$

$= 2\sqrt{49} + 2\sqrt{21} - 3\sqrt{21} - 3\sqrt{9}$

$= 14 - \sqrt{21} - 9$

$= 5 - \sqrt{21}$

31. $\left(\sqrt{3}+2\sqrt{6}\right)\left(2\sqrt{3}-\sqrt{6}\right)$

$= 2\sqrt{9}-\sqrt{18}+4\sqrt{18}-2\sqrt{36}$

$= 6+3\sqrt{18}-12$

$= -6+3\sqrt{9\cdot2}$

$= -6+9\sqrt{2}$

33. $\left(3\sqrt{7}-\sqrt{8}\right)\left(\sqrt{8}+2\sqrt{7}\right)$

$= 3\sqrt{56}+6\sqrt{49}-\sqrt{64}-2\sqrt{56}$

$= \sqrt{56}+42-8$

$= \sqrt{4\cdot14}+34$

$= 2\sqrt{14}+34$

35. $\left(2\sqrt{5}-3\right)^2 = 4\sqrt{25}+2\left(-6\sqrt{5}\right)+9$

$= 20-12\sqrt{5}+9$

$= 29-12\sqrt{5}$

37. $\left(\sqrt{3}+5\sqrt{2}\right)^2 = \sqrt{9}+2\left(5\sqrt{6}\right)+25\sqrt{4}$

$= 3+10\sqrt{6}+50$

$= 53+10\sqrt{6}$

39. $\left(\sqrt{6}-2\sqrt{3}\right)^2 = \sqrt{36}+2\left(-2\sqrt{18}\right)+4\sqrt{9}$

$= 6-4\sqrt{18}+12$

$= 18-4\sqrt{9\cdot2}$

$= 18-12\sqrt{2}$

41. $\left(\sqrt{a}-\sqrt{ab}\right)^2 = \sqrt{a^2}+2\left(-\sqrt{a^2b}\right)+\sqrt{a^2b^2}$

$= a-2a\sqrt{b}+ab$

43. $\left(2a\sqrt{3}-\sqrt{a}\right)\left(\sqrt{3}+\sqrt{a}\right)$

$= 2a\sqrt{9}+2a\sqrt{3a}-\sqrt{3a}-\sqrt{a^2}$

$= 6a+2a\sqrt{3a}-\sqrt{3a}-a$

$= 5a+2a\sqrt{3a}-\sqrt{3a}$

45. $\left(5x\sqrt{x}-2\sqrt{5x}\right)\left(\sqrt{5}-\sqrt{x}\right)$

$= 5x\sqrt{5x}-5x\sqrt{x^2}-2\sqrt{25x}+2\sqrt{5x^2}$

$= 5x\sqrt{5x}-5x^2-10\sqrt{x}+2x\sqrt{5}$

47. (a) $\sqrt{a}\cdot\sqrt{a}=a$ is not true when a is negative.

(b) The restriction "for any nonnegative real number" is necessary in order to deal with real numbers and it is not true when 'a' is negative.

Cumulative Review Problems

49. $36x^2-49y^2 = \left(6x\right)^2-\left(7y\right)^2$

$= \left(6x+7y\right)\left(6x-7y\right)$

51. $x =$ amount invested at 6%

$40{,}000-x =$ amount invested at 11%

$0.06x+0.11\left(40{,}000-x\right)=3775$

$6x+11\left(40{,}000-x\right)=377{,}500$

$6x+440{,}000-11x=377{,}500$

$-5x=-62{,}500$

$x=12{,}500$ and

$40{,}000-x=27{,}500$

She invested \$12,500 at 6% and \$27,500 at 11%.

Exercises 9.5

1. $\dfrac{\sqrt{32}}{\sqrt{2}}=\sqrt{\dfrac{32}{2}}=\sqrt{16}=4$

3. $\dfrac{\sqrt{5}}{\sqrt{20}}=\sqrt{\dfrac{5}{20}}=\sqrt{\dfrac{1}{4}}=\dfrac{1}{2}$

5. $\dfrac{\sqrt{27}}{\sqrt{3}}=\sqrt{\dfrac{27}{3}}=\sqrt{9}=3$

7. $\dfrac{\sqrt{6}}{\sqrt{x^4}} = \dfrac{\sqrt{6}}{x^2}$

9. $\dfrac{\sqrt{75}}{\sqrt{3}} = \sqrt{\dfrac{75}{3}} = \sqrt{25} = 5$

11. $\dfrac{\sqrt{8}}{\sqrt{x^6}} = \dfrac{\sqrt{4\cdot 2}}{x^3} = \dfrac{2\sqrt{2}}{x^3}$

13. $\dfrac{3}{\sqrt{7}} = \dfrac{3}{\sqrt{7}}\cdot\dfrac{\sqrt{7}}{\sqrt{7}} = \dfrac{3\sqrt{7}}{\sqrt{49}} = \dfrac{3\sqrt{7}}{7}$

15. $\dfrac{5}{\sqrt{6}} = \dfrac{5}{\sqrt{6}}\cdot\dfrac{\sqrt{6}}{\sqrt{6}} = \dfrac{5\sqrt{6}}{\sqrt{36}} = \dfrac{5\sqrt{6}}{6}$

17. $\dfrac{x}{\sqrt{3}} = \dfrac{x}{\sqrt{3}}\cdot\dfrac{\sqrt{3}}{\sqrt{3}} = \dfrac{x\sqrt{3}}{\sqrt{9}} = \dfrac{x\sqrt{3}}{3}$

19. $\dfrac{\sqrt{8}}{\sqrt{x}} = \dfrac{\sqrt{4\cdot 2}}{\sqrt{x}}\cdot\dfrac{\sqrt{x}}{\sqrt{x}} = \dfrac{2\sqrt{2x}}{\sqrt{x^2}} = \dfrac{2\sqrt{2x}}{x}$

21. $\dfrac{3}{\sqrt{12}} = \dfrac{3}{\sqrt{4\cdot 3}} = \dfrac{3}{2\sqrt{3}} = \dfrac{3}{2\sqrt{3}}\cdot\dfrac{\sqrt{3}}{\sqrt{3}} = \dfrac{3\sqrt{3}}{2\sqrt{9}} = \dfrac{\sqrt{3}}{2}$

23. $\dfrac{7}{\sqrt{a^3}} = \dfrac{7}{\sqrt{a^2 a}} = \dfrac{7}{a\sqrt{a}} = \dfrac{7}{a\sqrt{a}}\cdot\dfrac{\sqrt{a}}{\sqrt{a}} = \dfrac{7\sqrt{a}}{a\sqrt{a^2}}$
$= \dfrac{7\sqrt{a}}{a^2}$

25. $\dfrac{x}{\sqrt{2x^5}} = \dfrac{x}{\sqrt{2xx^4}} = \dfrac{x}{x^2\sqrt{2x}} = \dfrac{1}{x\sqrt{2x}}\cdot\dfrac{\sqrt{2x}}{\sqrt{2x}}$
$= \dfrac{\sqrt{2x}}{x\sqrt{4x^2}} = \dfrac{\sqrt{2x}}{2x^2}$

27. $\dfrac{\sqrt{3}}{\sqrt{12x^7}} = \dfrac{\sqrt{3}}{\sqrt{4\cdot 3xx^6}} = \dfrac{\sqrt{3}}{2x^3\sqrt{3x}} = \dfrac{1}{2x^3}\sqrt{\dfrac{3}{3x}}$
$= \dfrac{1}{2x^3}\cdot\dfrac{1}{\sqrt{x}} = \dfrac{1}{2x^3\sqrt{x}}\cdot\dfrac{\sqrt{x}}{\sqrt{x}} = \dfrac{\sqrt{x}}{2x^3\sqrt{x^2}}$
$= \dfrac{\sqrt{x}}{2x^4}$

29. $\sqrt{\dfrac{2}{7}} = \dfrac{\sqrt{2}}{\sqrt{7}}\cdot\dfrac{\sqrt{7}}{\sqrt{7}} = \dfrac{\sqrt{14}}{\sqrt{49}} = \dfrac{\sqrt{14}}{7}$

31. $\sqrt{\dfrac{3}{ab^2}} = \dfrac{\sqrt{3}}{\sqrt{ab^2}} = \dfrac{\sqrt{3}}{b\sqrt{a}}\cdot\dfrac{\sqrt{a}}{\sqrt{a}} = \dfrac{\sqrt{3a}}{b\sqrt{a^2}} = \dfrac{\sqrt{3a}}{ba}$

33. $\dfrac{9}{\sqrt{32x}} = \dfrac{9}{\sqrt{16\cdot 2x}} = \dfrac{9}{4\sqrt{2x}}\cdot\dfrac{\sqrt{2x}}{\sqrt{2x}} = \dfrac{9\sqrt{2x}}{4\sqrt{4x^2}}$
$= \dfrac{9\sqrt{2x}}{8x}$

35. $\dfrac{3}{\sqrt{2}-1} = \dfrac{3}{\sqrt{2}-1}\cdot\dfrac{\sqrt{2}+1}{\sqrt{2}+1}$
$= \dfrac{3\sqrt{2}+3}{\left(\sqrt{2}\right)^2-(1)^2}$
$= 3\sqrt{2}+3$

37. $\dfrac{3}{\sqrt{2}+\sqrt{5}} = \dfrac{3}{\sqrt{2}+\sqrt{5}}\cdot\dfrac{\sqrt{2}-\sqrt{5}}{\sqrt{2}-\sqrt{5}}$
$= \dfrac{3\sqrt{2}-3\sqrt{5}}{\left(\sqrt{2}\right)^2-\left(\sqrt{5}\right)^2}$
$= \dfrac{3\sqrt{2}-3\sqrt{5}}{-3}$
$= \sqrt{5}-\sqrt{2}$

39. $\dfrac{\sqrt{6}}{\sqrt{6}-\sqrt{3}} = \dfrac{\sqrt{6}}{\sqrt{6}-\sqrt{3}}\cdot\dfrac{\sqrt{6}+\sqrt{3}}{\sqrt{6}+\sqrt{3}}$
$= \dfrac{\sqrt{36}+\sqrt{18}}{\left(\sqrt{6}\right)^2-\left(\sqrt{3}\right)^2}$
$= \dfrac{6+\sqrt{9\cdot 2}}{6-3}$
$= \dfrac{6+3\sqrt{2}}{3}$
$= 2+\sqrt{2}$

41. $\dfrac{x}{\sqrt{7}+2\sqrt{2}} = \dfrac{x}{\sqrt{7}+2\sqrt{2}} \cdot \dfrac{\sqrt{7}-2\sqrt{2}}{\sqrt{7}-2\sqrt{2}}$

$\qquad = \dfrac{x\sqrt{7}-2x\sqrt{2}}{\left(\sqrt{7}\right)^2 - \left(2\sqrt{2}\right)^2}$

$\qquad = \dfrac{x\sqrt{7}-2x\sqrt{2}}{7-8}$

$\qquad = 2x\sqrt{2}-x\sqrt{7}$

43. $\dfrac{x}{\sqrt{3}-2x} = \dfrac{x}{\sqrt{3}-2x} \cdot \dfrac{\sqrt{3}+2x}{\sqrt{3}+2x}$

$\qquad = \dfrac{x\sqrt{3}+2x^2}{\left(\sqrt{3}\right)^2 - (2x)^2}$

$\qquad = \dfrac{x\sqrt{3}+2x^2}{3-4x^2}$

45. $\dfrac{\sqrt{7}}{\sqrt{8}+\sqrt{7}} = \dfrac{\sqrt{7}}{\sqrt{8}+\sqrt{7}} \cdot \dfrac{\sqrt{8}-\sqrt{7}}{\sqrt{8}-\sqrt{7}}$

$\qquad = \dfrac{\sqrt{56}-\sqrt{49}}{\left(\sqrt{8}\right)^2 - \left(\sqrt{7}\right)^2}$

$\qquad = \dfrac{\sqrt{4\cdot 14}-7}{8-7}$

$\qquad = 2\sqrt{14}-7$

47. $\dfrac{\sqrt{5}-\sqrt{2}}{\sqrt{5}+\sqrt{2}} = \dfrac{\sqrt{5}-\sqrt{2}}{\sqrt{5}+\sqrt{2}} \cdot \dfrac{\sqrt{5}-\sqrt{2}}{\sqrt{5}-\sqrt{2}}$

$\qquad = \dfrac{\sqrt{25}+2\left(-\sqrt{10}\right)+\sqrt{4}}{\left(\sqrt{5}\right)^2 - \left(\sqrt{2}\right)^2}$

$\qquad = \dfrac{5-2\sqrt{10}+2}{5-2}$

$\qquad = \dfrac{7-2\sqrt{10}}{3}$

49. $\dfrac{2\sqrt{3}+\sqrt{6}}{\sqrt{6}-\sqrt{3}} = \dfrac{2\sqrt{3}+\sqrt{6}}{\sqrt{6}-\sqrt{3}} \cdot \dfrac{\sqrt{6}+\sqrt{3}}{\sqrt{6}+\sqrt{3}}$

$\qquad = \dfrac{2\sqrt{18}+2\sqrt{9}+\sqrt{36}+\sqrt{18}}{\left(\sqrt{6}\right)^2 - \left(\sqrt{3}\right)^2}$

$\qquad = \dfrac{2\sqrt{9\cdot 2}+6+6+\sqrt{9\cdot 2}}{6-3}$

$\qquad = \dfrac{6\sqrt{2}+12+3\sqrt{2}}{3}$

$\qquad = 3\sqrt{2}+4$

51. $\dfrac{3\sqrt{5}+6}{\sqrt{7}-\sqrt{8}} = \dfrac{3\sqrt{5}+6}{\sqrt{7}-\sqrt{8}} \cdot \dfrac{\sqrt{7}+\sqrt{8}}{\sqrt{7}+\sqrt{8}}$

$\qquad = \dfrac{3\sqrt{35}+3\sqrt{40}+6\sqrt{7}+6\sqrt{8}}{\left(\sqrt{7}\right)^2 - \left(\sqrt{8}\right)^2}$

$\qquad = \dfrac{3\sqrt{35}+3\sqrt{4\cdot 10}+6\sqrt{7}+6\sqrt{4\cdot 2}}{7-8}$

$\qquad = -3\sqrt{35}-6\sqrt{10}-6\sqrt{7}-12\sqrt{2}$

53. $\dfrac{x-4}{\sqrt{x}+2} = \dfrac{x-4}{\sqrt{x}+2} \cdot \dfrac{\sqrt{x}-2}{\sqrt{x}-2}$

$\qquad = \dfrac{(x-4)(\sqrt{x}-2)}{\left(\sqrt{x}\right)^2 - (2)^2}$

$\qquad = \dfrac{(x-4)(\sqrt{x}-2)}{x-4}$

$\qquad = \sqrt{x}-2$

55. $\dfrac{\sqrt{x}+2}{\sqrt{x}-\sqrt{2}} = \dfrac{\sqrt{x}+2}{\sqrt{x}-\sqrt{2}} \cdot \dfrac{\sqrt{x}+\sqrt{2}}{\sqrt{x}+\sqrt{2}}$

$\qquad = \dfrac{\sqrt{x^2}+\sqrt{2x}+2\sqrt{x}+2\sqrt{2}}{\left(\sqrt{x}\right)^2 - \left(\sqrt{2}\right)^2}$

$\qquad = \dfrac{x+\sqrt{2x}+2\sqrt{x}+2\sqrt{2}}{x-2}$

To Think About

57. (a) $(x+\sqrt{2})(x-\sqrt{2})=(x)^2-(\sqrt{2})^2=x^2-2$

(b) $x^2-2=(x)^2-(\sqrt{2})^2=(x+\sqrt{2})(x-\sqrt{2})$

(c) $x^2-12=(x)^2-(2\sqrt{3})^2=(x+2\sqrt{3})(x-2\sqrt{3})$

59. $\dfrac{\sqrt{7}}{\sqrt{5}}=1.183215957$

$\dfrac{\sqrt{35}}{5}=1.183215957$

The values are the same.
They are equivalent.

Cumulative Review Problems

61. If $x=12$ and $y=-3$

$3x^2-x+y+y^2-y$

$=3(12)^2-(12)+(-3)+(-3)^2-(-3)$

$=3(144)-12+(-3)+9-(-3)$

$=432+4+9+3$

$=448$

63. $x=$ score on final

$\dfrac{70+70+60+90+80+x}{7}=80$

$370+x=560$

$x=190$ points

Exercises 9.6

1. $3^2+4^2=c^2$

$9+16=c^2$

$25=c^2$

$c=5$

3. $12^2+b^2=13^2$

$144+b^2=169$

$b^2=25$

$b=5$

5. $a=5,\ b=7$

$5^2+7^2=c^2$

$25+49=c^2$

$74=c^2$

$c=\sqrt{74}$

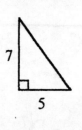

7. $a=\sqrt{18},\ b=4$

$(\sqrt{18})^2+4^2=c^2$

$18+16=c^2$

$34=c^2$

$c=\sqrt{34}$

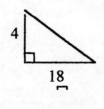

9. $c=20,\ b=18$

$a^2+18^2=20^2$

$a^2+324=400$

$a^2=76$

$a=\sqrt{76}=2\sqrt{19}$

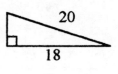

11. $c=\sqrt{23},\ a=4$

$4^2+b^2=(\sqrt{23})^2$

$16+b^2=23$

$b^2=7$

$b=\sqrt{7}$

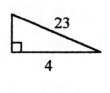

13. $c=12.96,\ b=8.35$

$a^2+(8.35)^2=(12.96)^2$

$a^2+69.7225=167.9616$

$a^2=98.2391$

$a\approx9.91$

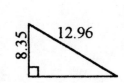

15.

15. continued

$$15^2 + x^2 = 18^2$$
$$225 + x^2 = 324$$
$$x^2 = 99$$
$$x = \sqrt{99} \approx 9.9$$

The base is 9.9 ft from the wall.

17.

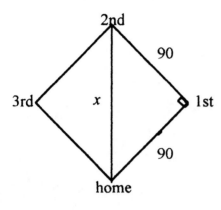

$$90^2 + 90^2 = x^2$$
$$8100 + 8100 = x^2$$
$$16{,}200 = x^2$$
$$x = \sqrt{16{,}200} \approx 127.3$$

It is 127.3 feet from home plate to second base.

19.
$$3^2 + 6^2 = x^2$$
$$9 + 36 = x^2$$
$$45 = x^2$$
$$x = \sqrt{45} \approx 6.7$$

He is 6.7 miles from the top of the mountain.

21. $\sqrt{x} = 11$ Check: $\sqrt{121} \overset{?}{=} 11$
$$x = 11^2 \qquad\qquad 11 = 11$$
$$x = 121$$

23. $\sqrt{x+8} = 5$ Check: $\sqrt{17+8} \overset{?}{=} 5$
$$x + 8 = 5^2 \qquad\qquad \sqrt{25} \overset{?}{=} 5$$
$$x + 8 = 25 \qquad\qquad 5 = 5$$
$$x = 17$$

25. $\sqrt{3x+6} = 2$ Check: $\sqrt{3\left(-\dfrac{2}{3}\right)+6} \overset{?}{=} 2$
$$3x + 6 = 2^2 \qquad\qquad \sqrt{-2+6} \overset{?}{=} 2$$
$$3x + 6 = 4 \qquad\qquad \sqrt{4} \overset{?}{=} 2$$
$$3x = -2 \qquad\qquad 2 = 2$$
$$x = -\frac{2}{3}$$

27. $\sqrt{2x+2} = \sqrt{3x-5}$
$$2x + 2 = 3x - 5$$
$$2 = x - 5$$
$$7 = x$$

Check: $\sqrt{2(7)+2} \overset{?}{=} \sqrt{3(7)-5}$
$$\sqrt{16} \overset{?}{=} \sqrt{16}$$
$$4 = 4$$

29. $\sqrt{2x} - 5 = 4$ Check: $\sqrt{2\left(\dfrac{81}{2}\right)} - 5 \overset{?}{=} 4$
$$\sqrt{2x} = 9 \qquad\qquad \sqrt{81} - 5 \overset{?}{=} 4$$
$$2x = 9^2 \qquad\qquad 9 - 5 \overset{?}{=} 4$$
$$2x = 81 \qquad\qquad 4 = 4$$
$$x = \frac{81}{2}$$

31. $\sqrt{3x+10} = x$
$$3x + 10 = x^2$$
$$x^2 - 3x - 10 = 0$$
$$(x - 5)(x + 2) = 0$$
$$x - 5 = 0 \qquad x + 2 = 0$$
$$x = 5 \qquad\quad x = -2$$

31. continued

Check:

$$\sqrt{3(5)+10} \stackrel{?}{=} 5 \qquad \sqrt{3(-2)+10} \stackrel{?}{=} -2$$

$$\sqrt{15+10} \stackrel{?}{=} 5 \qquad \sqrt{-6+10} \stackrel{?}{=} -2$$

$$\sqrt{25} \stackrel{?}{=} 5 \qquad \sqrt{4} \stackrel{?}{=} -2$$

$$5 = 5 \qquad\qquad 2 \neq -2$$

$$x = 5 \text{ only}$$

33. $\sqrt{5y+1} = y+1$

$$5y+1 = (y+1)^2$$

$$5y+1 = y^2 + 2y + 1$$

$$y^2 - 3y = 0$$

$$y(y-3) = 0$$

$$y = 0, \quad y - 3 = 0$$

$$y = 3$$

Check:

$$\sqrt{5(0)+1} \stackrel{?}{=} 0+1 \qquad \sqrt{5(3)+1} \stackrel{?}{=} 3+1$$

$$\sqrt{1} = 1 \qquad\qquad \sqrt{16} \stackrel{?}{=} 4$$

$$1 = 1 \qquad\qquad 4 = 4$$

$$y = 0, 3$$

35. $\qquad x - 3 = \sqrt{2x-3}$

$$(x-3)^2 = 2x-3$$

$$x^2 - 6x + 9 = 2x - 3$$

$$x^2 - 8x + 12 = 0$$

$$(x-6)(x-2) = 0$$

$$x - 6 = 0 \qquad x - 2 = 0$$

$$x = 6 \qquad\qquad x = 2$$

35. continued

Check:

$$6 - 3 \stackrel{?}{=} \sqrt{2(6)-3} \qquad 2 - 3 \stackrel{?}{=} \sqrt{2(2)-3}$$

$$3 \stackrel{?}{=} \sqrt{9} \qquad\qquad -1 \stackrel{?}{=} \sqrt{1}$$

$$3 = 3 \qquad\qquad -1 \neq 1$$

$$x = 6 \text{ only}$$

37. $\sqrt{3y+1} - y = 1$

$$\sqrt{3y+1} = y+1$$

$$3y+1 = (y+1)^2$$

$$3y+1 = y^2 + 2y + 1$$

$$y^2 - y = 0$$

$$y(y-1) = 0$$

$$y = 0 \qquad y - 1 = 0$$

$$y = 1$$

Check:

$$\sqrt{3(0)+1} - 0 \stackrel{?}{=} 1 \qquad \sqrt{3(1)+1} - 1 \stackrel{?}{=} 1$$

$$\sqrt{1} - 0 \stackrel{?}{=} 1 \qquad\qquad \sqrt{4} - 1 \stackrel{?}{=} 1$$

$$1 = 1 \qquad\qquad 2 - 1 \stackrel{?}{=} 1$$

$$1 = 1$$

$$y = 0, 1$$

39. $\sqrt{5+2x} - 1 - x = 0$

$$\sqrt{5+2x} = x+1$$

$$5+2x = (x+1)^2$$

$$5+2x = x^2 + 2x + 1$$

$$x^2 - 4 = 0$$

$$(x+2)(x-2) = 0$$

$$x + 2 = 0 \qquad x - 2 = 0$$

$$x = -2 \qquad\qquad x = 2$$

176

39. continued

Check:

$$\sqrt{5+2(-2)}-1-(-2)\overset{?}{=}0 \qquad \sqrt{5+2(2)}-1-2\overset{?}{=}0$$

$$\sqrt{1}-1+2\overset{?}{=}0 \qquad \sqrt{9}-1-2\overset{?}{=}0$$

$$1-1+2\overset{?}{=}0 \qquad 3-1-2\overset{?}{=}0$$

$$2\neq0 \qquad\qquad 0=0$$

$$x=2 \text{ only}$$

41. $y+\sqrt{3y-14}=6$

$$\sqrt{3y-14}=6-y$$

$$3y-14=(6-y)^2$$

$$3y-14=36-12y+y^2$$

$$y^2-15y+50=0$$

$$(y-5)(y-10)=0$$

$$y-5=0 \qquad y-10=0$$

$$y=5 \qquad\qquad y=10$$

Check:

$$5+\sqrt{3(5)-14}\overset{?}{=}6 \qquad 10+\sqrt{3(10)-14}\overset{?}{=}6$$

$$5+\sqrt{1}\overset{?}{=}6 \qquad 10+\sqrt{16}\overset{?}{=}6$$

$$5+1\overset{?}{=}6 \qquad 10+4\overset{?}{=}6$$

$$6=6 \qquad\qquad 14\neq6$$

$$y=5 \text{ only}$$

43. $\sqrt{2x+5}=2\sqrt{2x}+1$

$$2x+5=\left(2\sqrt{2x}+1\right)^2$$

$$2x+5=8x+4\sqrt{2x}+1$$

$$-6x+4=4\sqrt{2x}$$

$$-3x+2=2\sqrt{2x}$$

$$(-3x+2)^2=8x$$

$$9x^2-12x+4=8x$$

$$9x^2-20x+4=0$$

43. continued

$$9x-2=0 \qquad x-2=0$$

$$9x=2 \qquad\quad x=2$$

$$x=\frac{2}{9}$$

Check:

$$\sqrt{2\left(\frac{2}{9}\right)+5}\overset{?}{=}2\sqrt{2\left(\frac{2}{9}\right)}+1 \qquad \sqrt{2(2)+5}\overset{?}{=}2\sqrt{2(2)}+1$$

$$\sqrt{\frac{49}{9}}\overset{?}{=}2\left(\sqrt{\frac{4}{9}}\right)+1 \qquad \sqrt{9}\overset{?}{=}2\sqrt{4}+1$$

$$\frac{7}{3}\overset{?}{=}2\left(\frac{2}{3}\right)+1 \qquad 3\overset{?}{=}2(2)+1$$

$$\frac{7}{3}\overset{?}{=}\frac{4}{3}+1 \qquad 3\overset{?}{=}4+1$$

$$\frac{7}{3}=\frac{7}{3} \qquad\qquad 3\neq5$$

$$x=\frac{2}{9} \text{ only}$$

<u>Cumulative Review Problems</u>

45.
$$\frac{2x}{x-7}+3=\frac{14}{x-7}$$

$$(x-7)\left(\frac{2x}{x-7}\right)+(x-7)(3)=(x-7)\left(\frac{14}{x-7}\right)$$

$$2x+3x-21=14$$

$$5x-21=14$$

$$5x=35$$

$$x=7 \text{ is extraneous}$$

No solution.

47. $f(x)=2x^2-3x+6$

$$f(-2)=2(-2)^2-3(-2)+6=20$$

1. $y = kx$, $y = 7$, $x = 4$

$7 = 4k \Rightarrow k = \dfrac{7}{4}$

$y = \dfrac{7}{4}x$, $x = 9$

$y = \dfrac{7}{4}(9) = \dfrac{63}{4}$

3. $y = kx$, $y = 350$, $x = 100$

$350 = 100k \Rightarrow k = \dfrac{7}{2}$

$y = \dfrac{7}{2}x$, $x = 20$

$y = \dfrac{7}{2}(20) = 70$

5. $y = kx^3$, $y = 12$, $x = 2$

$12 = k(2)^3 \Rightarrow k = \dfrac{12}{8} = \dfrac{3}{2}$

$y = \dfrac{3}{2}x^3$, $x = 7$

$y = \dfrac{3}{2}(7)^3 = \dfrac{3}{2}(343) = \dfrac{1029}{2}$

7. $y = kx^2$, $y = 800$, $x = 40$

$800 = k(40)^2 \Rightarrow k = \dfrac{1}{2}$

$y = \dfrac{1}{2}x^2$, $x = 25$

$y = \dfrac{1}{2}(25)^2 = \dfrac{625}{2}$

9. $y = kx^2$, $y = 1146.88$, $x = 12.80$

$1146.88 = k(12.80)^2 \Rightarrow k = 7$

$y = 7x^2$, $x = 7.30$

$y = 7(7.30)^2 = 373.03$

11. $p = $ pressure, $d = $ depth

$p = kd$, $p = 26$, $d = 60$

$26 = 60k \Rightarrow k = \dfrac{13}{30}$

$p = \dfrac{13}{30}d$, $d = 390$

$p = \dfrac{13}{30}(390) = 169 \dfrac{\text{lb}}{\text{sq. in}}$

13. $T = $ time, $s = $ length of side

$T = ks^3$, $T = 7$, $s = 2.0$

$7 = k(2.0)^3 \Rightarrow k = \dfrac{7}{8}$

$T = \dfrac{7}{8}s^3$, $s = 4.0$

$T = \dfrac{7}{8}(4.0)^3 = 56$ min.

15. $y = \dfrac{k}{x}$, $y = 12$, $x = 4$

$12 = \dfrac{k}{4} \Rightarrow k = 48$

$y = \dfrac{48}{x}$, $x = 7$

$y = \dfrac{48}{7}$

17. $y = \dfrac{k}{x}$, $y = \dfrac{1}{4}$, $x = 8$

$\dfrac{1}{4} = \dfrac{k}{8} \Rightarrow k = 2$

$y = \dfrac{2}{x}$, $x = 1$

$y = \dfrac{2}{1} = 2$

19. $y = \dfrac{k}{x^2}, \quad y = 100, \quad x = 5$

$100 = \dfrac{k}{(5)^2} \Rightarrow k = 2500$

$y = \dfrac{2500}{x^2}, \quad x = 3$

$y = \dfrac{2500}{(3)^2} = \dfrac{2500}{9}$

21. $y = \dfrac{k}{\sqrt{x}}, \quad y = 2, \quad x = 4$

$2 = \dfrac{k}{\sqrt{4}} \Rightarrow k = 4$

$y = \dfrac{4}{\sqrt{x}}, \quad x = 81$

$y = \dfrac{4}{\sqrt{81}} = \dfrac{4}{9}$

23. $y = \dfrac{k}{\sqrt{x}}, \quad y = 5.00, \quad x = 73.96$

$5.00 = \dfrac{k}{\sqrt{73.96}} \Rightarrow k = (8.6)(5.00) = 43$

$y = \dfrac{43}{\sqrt{x}}, \quad x = 18.49$

$y = \dfrac{43}{\sqrt{18.49}} = \dfrac{43}{4.3} = 10$

25. $C = $ current, $R = $ resistance

$C = \dfrac{k}{R}, \quad C = 42, \quad R = 5$

$42 = \dfrac{k}{5} \Rightarrow k = 210$

$C = \dfrac{210}{R}, \quad R = 4$

$C = \dfrac{210}{4} = \dfrac{105}{2} = 52.5$ amperes

27. $W = $ weight, $D = $ distance from center

$W = \dfrac{k}{D^2}, \quad W = 1000, \quad D = 4000$

$1000 = \dfrac{k}{(4000)^2} \Rightarrow k = 16{,}000{,}000{,}000$

$W = \dfrac{16{,}000{,}000{,}000}{D^2}, \quad D = 6000$

$W = \dfrac{16{,}000{,}000{,}000}{(6000)^2} = \dfrac{4000}{9} = 444\dfrac{4}{9}$ pounds

29. The graphs of $y = kx$ will all be straight lines through the origin.

31. $H = $ heat transfer, $A = $ area, $T = $ thickness
$D = $ temperature difference

$H = \dfrac{kAD}{T}, \quad A = 80, \quad D = 30, \quad T = 0.5, \quad H = 6000$

$6000 = \dfrac{k(80)(30)}{0.5} \Rightarrow k = 1.25$

$H = \dfrac{1.25\,AD}{T}, \quad A = 80, \quad D = 50, \quad T = 1$

$H = \dfrac{1.25(80)(50)}{1} = 5000$ BTU / hr

Cumulative Review Problems

33. $\dfrac{80{,}000}{320} = \dfrac{120{,}000}{320 + a}$

$\dfrac{2}{320} = \dfrac{3}{320 + a}$

$2(320 + a) = 3(320)$

$640 + 2a = 960$

$2a = 320$

$a = 160$

35. $12x^2 - 20x - 48,$

    $= 4(3x^2 - 5x - 12),$ Grouping number $= 36$

    $= 4(3x^2 - 9x + 4x - 12)$

    $= 4[3x(x - 3) + 4(x - 3)]$

    $= 4(x - 3)(3x + 4)$

Chapter 9 - Review Problems

1. $\sqrt{9} = 3$

3. $\sqrt{36} = 6$

5. $\sqrt{100} = 10$

7. $\sqrt{-81}$ is not a real number

9. $\sqrt{16} = 4$

11. $\sqrt{225} = 15$

13. $\sqrt{0.81} = 0.9$

15. $\sqrt{\dfrac{1}{25}} = \dfrac{1}{5}$

17. $-\sqrt{0.0004} = -0.02$

19. $\sqrt{-0.0016}$ is not a real number.

21. $\sqrt{105} = 10.247$

23. $\sqrt{77} = 8.775$

25. $\sqrt{50} = \sqrt{25 \cdot 2} = 5\sqrt{2}$

27. $\sqrt{98} = \sqrt{49 \cdot 2} = 7\sqrt{2}$

29. $\sqrt{40} = \sqrt{4 \cdot 10} = 2\sqrt{10}$

31. $\sqrt{x^8} = \sqrt{(x^4)^2} = x^4$

33. $\sqrt{x^5 y^6} = \sqrt{x^4 x y^6} = x^2 y^3 \sqrt{x}$

35. $\sqrt{16x^3 y^5} = \sqrt{16x^2 x y^4 y} = 4xy^2 \sqrt{xy}$

37. $\sqrt{12x^5} = \sqrt{4 \cdot 3x^4 x} = 2x^2 \sqrt{3x}$

39. $\sqrt{75x^{10}} = \sqrt{25 \cdot 3x^{10}} = 5x^5 \sqrt{3}$

41. $\sqrt{120a^3 b^4 c^5} = \sqrt{4 \cdot 30a^2 ab^4 c^4 c} = 2ab^2 c^2 \sqrt{30ac}$

43. $\sqrt{56x^7 y^9} = \sqrt{4 \cdot 14x^6 xy^8 y} = 2x^3 y^4 \sqrt{14xy}$

45. $\sqrt{2} - \sqrt{8} + \sqrt{32} = \sqrt{2} - \sqrt{4 \cdot 2} + \sqrt{16 \cdot 2}$

                            $= \sqrt{2} - 2\sqrt{2} + 4\sqrt{2}$

                            $= 3\sqrt{2}$

47. $x\sqrt{3} + 3x\sqrt{3} + \sqrt{27x^2} = 4x\sqrt{3} + \sqrt{9 \cdot 3x^2}$

                                 $= 4x\sqrt{3} + 3x\sqrt{3}$

                                 $= 7x\sqrt{3}$

49. $5\sqrt{5} - 6\sqrt{20} + 2\sqrt{10} = 5\sqrt{5} - 6\sqrt{4 \cdot 5} + 2\sqrt{10}$

                               $= 5\sqrt{5} - 12\sqrt{5} + 2\sqrt{10}$

                               $= -7\sqrt{5} + 2\sqrt{10}$

51. $2\sqrt{28} - 3\sqrt{63} + 2x\sqrt{7} = 2\sqrt{4 \cdot 7} - 3\sqrt{9 \cdot 7} + 2x\sqrt{7}$

                               $= 4\sqrt{7} - 9\sqrt{7} + 2x\sqrt{7}$

                               $= -5\sqrt{7} + 2x\sqrt{7}$

                               $= (2x - 5)\sqrt{7}$

53. $(2\sqrt{x})(3\sqrt{x^3}) = 6\sqrt{x^4} = 6x^2$

55. $(\sqrt{2a^3})(\sqrt{8b^2}) = \sqrt{16a^2 ab^2}$

                         $= 4ab\sqrt{a}$

57. $\sqrt{5}(3\sqrt{5} - \sqrt{20}) = \sqrt{5}(3\sqrt{5} - \sqrt{4 \cdot 5})$

                         $= \sqrt{5}(3\sqrt{5} - 2\sqrt{5})$

                         $= \sqrt{5}(\sqrt{5})$

                         $= 5$

59. $\sqrt{2}\left(\sqrt{5}-\sqrt{3}-2\sqrt{2}\right)=\sqrt{10}-\sqrt{6}-2\sqrt{4}$

$\qquad = \sqrt{10}-\sqrt{6}-4$

61. $\left(\sqrt{11}+2\right)\left(2\sqrt{11}-1\right)=2\sqrt{121}-\sqrt{11}+4\sqrt{11}-2$

$\qquad = 22+3\sqrt{11}-2$

$\qquad = 20+3\sqrt{11}$

63. $\left(2+3\sqrt{6}\right)\left(4-2\sqrt{3}\right)=8-4\sqrt{3}+12\sqrt{6}-6\sqrt{18}$

$\qquad = 8-4\sqrt{3}+12\sqrt{6}-18\sqrt{2}$

65. $\left(2\sqrt{3}+3\sqrt{6}\right)^{2}=4\sqrt{9}+2(6)\sqrt{18}+9\sqrt{36}$

$\qquad = 12+36\sqrt{2}+54$

$\qquad = 66+36\sqrt{2}$

67. $\left(a\sqrt{b}-2\sqrt{a}\right)\left(\sqrt{b}-3\sqrt{a}\right)$

$\quad = a\sqrt{b^{2}}-3a\sqrt{ab}-2\sqrt{ab}+6\sqrt{a^{2}}$

$\quad = ab-3a\sqrt{ab}-2\sqrt{ab}+6a$

69. $\dfrac{1}{\sqrt{3x}}=\dfrac{1}{\sqrt{3x}}\cdot\dfrac{\sqrt{3x}}{\sqrt{3x}}=\dfrac{\sqrt{3x}}{3x}$

71. $\dfrac{x^{2}y}{\sqrt{8}}=\dfrac{x^{2}y}{2\sqrt{2}}=\dfrac{x^{2}y}{2\sqrt{2}}\cdot\dfrac{\sqrt{2}}{\sqrt{2}}=\dfrac{x^{2}y\sqrt{2}}{4}$

73. $\sqrt{\dfrac{3}{7}}=\dfrac{\sqrt{3}}{\sqrt{7}}=\dfrac{\sqrt{3}}{\sqrt{7}}\cdot\dfrac{\sqrt{7}}{\sqrt{7}}=\dfrac{\sqrt{21}}{7}$

75. $\dfrac{15\sqrt{60}}{5\sqrt{15}}=\dfrac{15}{5}\sqrt{\dfrac{60}{15}}=3\sqrt{4}=6$

77. $\dfrac{\sqrt{a^{5}}}{\sqrt{2a}}=\dfrac{\sqrt{a^{5}}}{\sqrt{2a}}\cdot\dfrac{\sqrt{2a}}{\sqrt{2a}}=\dfrac{\sqrt{2a^{6}}}{2a}=\dfrac{a^{3}\sqrt{2}}{2a}=\dfrac{a^{2}\sqrt{2}}{2}$

79. $\dfrac{a\sqrt{a^{2}b}}{ab\sqrt{ab^{2}}}=\dfrac{a^{2}\sqrt{b}}{ab^{2}\sqrt{a}}=\dfrac{a\sqrt{b}}{b^{2}\sqrt{a}}\cdot\dfrac{\sqrt{a}}{\sqrt{a}}=\dfrac{a\sqrt{ab}}{b^{2}a}=\dfrac{\sqrt{ab}}{b^{2}}$

81. $\dfrac{3}{\sqrt{5}+\sqrt{2}}=\dfrac{3}{\sqrt{5}+\sqrt{2}}\cdot\dfrac{\sqrt{5}-\sqrt{2}}{\sqrt{5}-\sqrt{2}}=\dfrac{3\sqrt{5}-3\sqrt{2}}{\left(\sqrt{5}\right)^{2}-\left(\sqrt{2}\right)^{2}}$

$\qquad = \dfrac{3\sqrt{5}-3\sqrt{2}}{3}$

$\qquad = \sqrt{5}-\sqrt{2}$

83. $\dfrac{1-\sqrt{5}}{2+\sqrt{5}}=\dfrac{1-\sqrt{5}}{2+\sqrt{5}}\cdot\dfrac{2-\sqrt{5}}{2-\sqrt{5}}$

$\qquad = \dfrac{2-3\sqrt{5}+5}{2^{2}-\left(\sqrt{5}\right)^{2}}$

$\qquad = \dfrac{7-3\sqrt{5}}{-1}$

$\qquad = -7+3\sqrt{5}$

85. $c^{2}=a^{2}+b^{2},\quad a=5,\quad b=8$

$\quad c^{2}=5^{2}+8^{2}$

$\quad c=\sqrt{25+64}$

$\quad c=\sqrt{89}$

87. $\quad c^{2}=a^{2}+b^{2},\quad c=5,\quad a=3.5$

$\quad 5^{2}=(3.5)^{2}+b^{2}$

$\quad 25=12.25+b^{2}$

$\quad 12.75=b^{2}$

$\quad \sqrt{12\dfrac{3}{4}}=b$

$\quad \sqrt{\dfrac{51}{4}}=b$

$\quad \dfrac{\sqrt{51}}{2}=b$

89. $x^2 = 18^2 + 24^2$

$x^2 = 324 + 576$

$x^2 = 900$

$x = 30$

It is 30 meters from his feet to the top of the pole.

91. $10^2 = x^2 + x^2$

$100 = 2x^2$

$50 = x^2$

$7.1 = x$

Each side is 7.1 inches

93. $\sqrt{2x - 3} = 9$

$2x - 3 = 81$

$2x = 84$

$x = 42$

95. $\sqrt{-5 + 2x} = \sqrt{1 + x}$

$-5 + 2x = 1 + x$

$x = 6$

97. $\sqrt{2x - 5} = 10 - x$

$2x - 5 = (10 - x)^2$

$2x - 5 = 100 - 20x + x^2$

$0 = x^2 - 22x + 105$

$0 = (x - 7)(x - 15)$

$x = 15$ does not check

Solution: $x = 7$

99. $4x + \sqrt{x + 2} = 5x - 4$

$\sqrt{x + 2} = x - 4$

$x + 2 = (x - 4)^2$

$x + 2 = x^2 - 8x + 16$

$0 = x^2 - 9x + 14$

$0 = (x - 2)(x - 7)$

$x - 2 = 0 \quad$ or $\quad x - 7 = 0$

$x = 2 \qquad\qquad x = 7$

$x = 2$ does not check

Solution: $x = 7$

101. $y = \dfrac{k}{x^2}, \quad y = \dfrac{6}{5}, \quad x = 5$

$\dfrac{6}{5} = \dfrac{k}{5^2} \Rightarrow k = 30$

$y = \dfrac{30}{x^2}, \quad x = 15$

$y = \dfrac{30}{(15)^2} = \dfrac{2}{15}$

103. $P = $ population, $p = $ weight of pesticide

$P = \dfrac{k}{p}, \quad P = 1000, \quad p = 40$

$1000 = \dfrac{k}{40} \Rightarrow k = 40,000$

$P = \dfrac{40,000}{p}, \quad P = 100$

$100 = \dfrac{40,000}{p} \Rightarrow p = 400$

Need 400 lb of pesticide.

105. $H = $ horsepower, $s = $ speed

$H = ks^3, \quad s = s_m, \quad H = H_m$

$H_1 = k(s_m)^3 \Rightarrow k = \dfrac{H_1}{(s_m)^3}$

$H_1 = \dfrac{H_m}{s_m^3} s^3, \quad s = 2s_m$

$H = \dfrac{H_m}{s_m^3}(25_m)^3 = 8H_m$

The horsepower is 8 times as much.

Puting Your Skills To Work

Use $r = 2\sqrt{5L}$

1. $L = 5$

$r = 2\sqrt{5 \cdot 5} = 10$ miles per hour

3. $L = 125$

$r = 2\sqrt{5 \cdot 125} = 50$ miles per hour

5. $r = 20$

$20 = 2\sqrt{5L}$

$400 = 4(5L)$

$L = 20$ feet

7. $r = 80$

$80 = 2\sqrt{5L}$

$6400 = 4(5L)$

$L = 320$ feet

9.

L	0	50	100	200	300	400
r	0	31.6	44.7	63.2	77.5	89.4

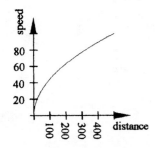

11. $r = 80, \quad L = \left(\dfrac{80}{2}\right)^2 / 5 = 320$ feet

$r = 100, \quad L = \left(\dfrac{100}{2}\right)^2 / 5 = 500$ feet

This is a 180-foot increase. This is 3 times the 60-foot increase that occurred when speeding up from 20 to 40 miles per hour.

13. The rate of speed could not be accurately calculated and a case against the driver may not hold up in a court of law.

1. $\sqrt{81} = 9$

3. $\sqrt{48x^2 y^7} = \sqrt{3 \cdot 16x^2 y^6 y}$

$\qquad = 4xy^3 \sqrt{3y}$

5. $\sqrt{5} + 3\sqrt{20} - 2\sqrt{45} = \sqrt{5} + 3\sqrt{4 \cdot 5} - 2\sqrt{9 \cdot 5}$

$\qquad\qquad = \sqrt{5} + 6\sqrt{5} - 6\sqrt{5}$

$\qquad\qquad = \sqrt{5}$

7. $(2\sqrt{a})(3\sqrt{b})(2\sqrt{ab}) = 12\sqrt{a^2 b^2}$

$\qquad\qquad\qquad\qquad = 12ab$

9. $(2 - \sqrt{6})^2 = 4 + 2(-2\sqrt{6}) + \sqrt{36}$

$\qquad\qquad = 4 - 4\sqrt{6} + 6$

$\qquad\qquad = 10 - 4\sqrt{6}$

11. $\sqrt{\dfrac{x}{5}} = \dfrac{\sqrt{x}}{\sqrt{5}} = \dfrac{\sqrt{x}}{\sqrt{5}} \cdot \dfrac{\sqrt{5}}{\sqrt{5}} = \dfrac{\sqrt{5x}}{5}$

13. $\dfrac{3 - \sqrt{2}}{\sqrt{2} - 6} = \dfrac{3 - \sqrt{2}}{\sqrt{2} - 6} \cdot \dfrac{\sqrt{2} + 6}{\sqrt{2} + 6}$

$\qquad = \dfrac{3\sqrt{2} + 18 - \sqrt{4} - 6\sqrt{2}}{(\sqrt{2})^2 - 6^2}$

$\qquad = \dfrac{16 - 3\sqrt{2}}{-34}$

15. $\sqrt{156} = 12.49$

17. $x^2 = (3\sqrt{2})^2 + 3^2 = 9\sqrt{4} + 9 = 27$

$x = \sqrt{27} = \sqrt{9 \cdot 3} = 3\sqrt{3}$

183

19.
$$x = 5 + \sqrt{x+7}$$
$$x - 5 = \sqrt{x+7}$$
$$x^2 - 10x + 25 = x + 7$$
$$x^2 - 11x + 18 = 0$$
$$(x-2)(x-9) = 0$$

$$x - 2 = 0 \qquad x - 9 = 0$$
$$x = 2 \qquad x = 9$$

2 is an extraneous solution. $x = 9$

21.
$$T = \text{tax}, \quad C = \text{cost}$$
$$T = kC, \quad T = 26.10, \quad C = 360$$
$$26.10 = k(360) \Rightarrow k = 0.0725$$
$$T = 0.0725C, \quad C = 12{,}580$$
$$T = 0.0725(12{,}580) = 912.05$$

The tax is \$912.05.

Cumulative Test - Chapters 0 - 9

1. $\dfrac{12}{15} = \dfrac{4 \cdot 3}{5 \cdot 3} = \dfrac{4}{5}$

3. $6\dfrac{1}{4} + 6\dfrac{2}{3} = \dfrac{25}{4} + \dfrac{20}{3}$

$\qquad = \dfrac{25}{4} \cdot \dfrac{3}{20}$

$\qquad = \dfrac{5(5)}{4} \cdot \dfrac{3}{4(5)}$

$\qquad = \dfrac{15}{16}$

5. $-11 - 16 + 8 + 4 - 13 + 31$
$\quad = -27 + 8 + 4 - 13 + 31$
$\quad = -19 + 4 - 13 + 31$
$\quad = -15 - 13 + 31$
$\quad = -28 + 31$
$\quad = 3$

7. $3x^3yz^2 + 4xyz - 5x^3yz^2$
$\quad = -2x^3yz^2 + 4xyz$

9. $(2x-3)^2 = 4x^2 + 2(-3)(2x) + 9$
$\qquad\qquad = 4x^2 - 12x + 9$

11. $\dfrac{4x}{x+2} + \dfrac{4}{x+2} = \dfrac{4x+4}{x+2}$

13. $\sqrt{98x^5y^6} = \sqrt{49 \cdot 2x^4xy^6}$
$\qquad\qquad = 7x^2y^3\sqrt{2x}$

15. $\left(\sqrt{6} - \sqrt{3}\right)^2 = \sqrt{36} + 2\left(-\sqrt{3}\right)\left(\sqrt{6}\right) + \sqrt{9}$
$\qquad\qquad = 6 - 2\sqrt{18} + 3$
$\qquad\qquad = 9 - 2\sqrt{9 \cdot 2}$
$\qquad\qquad = 9 - 6\sqrt{2}$

17. $\left(3\sqrt{6} - \sqrt{2}\right)\left(\sqrt{6} + 4\sqrt{2}\right)$
$\quad = 3\sqrt{36} + 12\sqrt{12} - \sqrt{12} - 4\sqrt{4}$
$\quad = 18 + 11\sqrt{12} - 8$
$\quad = 10 + 11\sqrt{4 \cdot 3}$
$\quad = 10 + 22\sqrt{3}$

19. $-\sqrt{16} = -4$

21. $c^2 = a^2 + b^2, \quad a = 4, b = \sqrt{7}$
$\quad c^2 = 4^2 + \left(\sqrt{7}\right)^2 = 16 + 7 = 23$
$\qquad c = \sqrt{23}$

23. $\sqrt{4x+5} = x$
$\qquad 4x + 5 = x^2$
$\qquad\quad 0 = x^2 - 4x - 5$
$\qquad\quad 0 = (x-5)(x+1)$

$\quad x - 5 = 0 \qquad x + 1 = 0$
$\qquad x = 5 \qquad\quad x = -1$

-1 is an extraneous solution. $x = 5$

25. $y = \dfrac{k}{x}$, $y = 4$, $x = 5$

$4 = \dfrac{k}{5} \Rightarrow k = 20$

$y = \dfrac{20}{x}$, $x = 2$

$y = \dfrac{20}{2} = 10$

Practice Quiz: Sections 9.1 - 9.3

1. Evaluate $-\sqrt{169}$

2. Evaluate $\sqrt{\dfrac{16}{49}}$

3. Simplify $\sqrt{12x^3 y^2}$

4. Combine $4\sqrt{3} - 2\sqrt{3}$

5. Combine $\sqrt{20} - \sqrt{125}$

Practice Quiz: Sections 9.4 - 9.7

Multiply and simplify your answer in problems 1 and 2.

1. $\left(3\sqrt{2}\right)\left(5\sqrt{10}\right)$

2. $\left(\sqrt{3} + \sqrt{7}\right)\left(2\sqrt{3} - \sqrt{7}\right)$

3. Rationalize the denominator

$\dfrac{4}{2 - \sqrt{3}}$

4. Solve $\sqrt{2x + 1} = x - 1$

5. If y varies directly as the square of x and $y = \dfrac{9}{4}$ when $x = 3$, find the value of y if $x = 2$.

Answers to Practice Quiz

Sections 9.1 - 9.3

1. -13

2. $\dfrac{4}{7}$

3. $2xy\sqrt{3x}$

4. $2\sqrt{3}$

5. $-3\sqrt{5}$

Answers to Practice Quiz

Sections 9.4 - 9.7

1. $30\sqrt{5}$

2. $-1 + \sqrt{21}$

3. $8 + 4\sqrt{3}$

4. $x = 4$

5. $y = 1$

1. $x^2 - 12x - 28 = 0$

$(x - 14)(x + 2) = 0$

$x - 14 = 0 \qquad x + 2 = 0$

$\qquad x = 14 \qquad\qquad x = -2$

3. $\qquad\qquad 5x^2 = 22x - 8$

$5x^2 - 22x + 8 = 0$

$(5x - 2)(x - 4) = 0$

$5x - 2 = 0 \qquad x - 4 = 0$

$\quad 5x = 2 \qquad\qquad x = 4$

$\quad x = \dfrac{2}{5}$

5. $x^2 - 12 = 0$

$\quad x^2 = 12$

$\quad x = \pm\sqrt{12}$

$\quad x = \pm\sqrt{4 \cdot 3}$

$\quad x = \pm 2\sqrt{3}$

7. $x^2 + 8x + 5 = 0$

$x^2 + 8x = -5, \quad \left[\dfrac{1}{2}(8)\right]^2 = 4^2 = 16$

$x^2 + 8x + 16 = -5 + 16$

$\quad (x + 4)^2 = 11$

$\quad\quad x + 4 = \pm\sqrt{11}$

$\quad\quad\quad x = -4 \pm \sqrt{11}$

9. $2x^2 - 6x + 3 = 0$

$a = 2, \ b = -6, \ c = 3$

$x = \dfrac{-(-6) \pm \sqrt{(-6)^2 - 4(2)(3)}}{2(2)}$

$\quad = \dfrac{6 \pm \sqrt{12}}{4}$

$\quad = \dfrac{6 \pm 2\sqrt{3}}{4}$

$\quad = \dfrac{3 \pm \sqrt{3}}{2}$

11. $3x^2 + 8x + 1 = 0$

$a = 3, \ b = 8, \ c = 1$

$x = \dfrac{-8 \pm \sqrt{8^2 - 4(3)(1)}}{2(3)}$

$\quad = \dfrac{-8 \pm \sqrt{52}}{6}$

$\quad = \dfrac{-8 \pm 2\sqrt{13}}{6}$

$\quad = \dfrac{-4 \pm \sqrt{13}}{3}$

13. $y = 5 - 3x^2$

Vertex: $x = -\dfrac{0}{2(-3)} = 0$

$\qquad\qquad y = 5 - 3(0)^2 = 5$

V$(0, 5)$

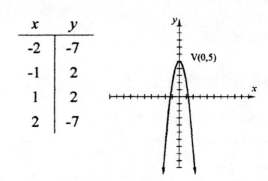

x	y
-2	-7
-1	2
1	2
2	-7

15. $\qquad x = $ length of base

$2x + 1 = $ altitude

$\text{Area} = \dfrac{1}{2}bh$

$\quad 39 = \dfrac{1}{2}x(2x + 1)$

$\quad 78 = 2x^2 + x$

$\quad\quad 0 = 2x^2 + x - 78$

$\quad\quad 0 = (2x + 13)(x - 6)$

$2x + 13 = 0 \qquad\qquad x - 6 = 0$

$\quad 2x = -13 \qquad\qquad\quad x = 6$

$\quad\quad x = -\dfrac{13}{2}$

15. continued

The length can't be negative,

so $x = 6$, $2x + 1 = 13$

Base $= 6$ cm, altitude $= 13$ cm.

<u>Exercises 10.1</u>

1. $x^2 + 5x + 6 = 0$

$a = 1$, $b = 5$, $c = 6$

3. $7x^2 - 11x - 2 = 0$

$a = 7$, $b = -11$, $c = -2$

5. $2 = -12x^2 + 3x$

$0 = -12x^2 + 3x - 2$

$a = -12$, $b = 3$, $c = -2$

7. $3x^2 - 8x = 0$

$a = 3$, $b = -8$, $c = 0$

9. $x^2 + 15x - 7 = 12x + 8$

$x^2 + 3x - 15 = 0$

$a = 1$, $b = 3$, $c = -15$

11. $\quad 2x(x - 3) = (x + 5)(x - 2)$

$\quad 2x^2 - 6x = x^2 + 3x - 10$

$x^2 - 9x + 10 = 0$

$a = 1$, $b = -9$, $c = 10$

13. $18x^2 + 9x = 0$

$9x(2x + 1) = 0$

$9x = 0 \qquad 2x + 1 = 0$

$x = 0 \qquad\quad 2x = -1$

$\qquad\qquad x = -\dfrac{1}{2}$

15. $\qquad 11x^2 = 14x$

$11x^2 - 14x = 0$

$x(11x - 14) = 0$

$x = 0 \qquad 11x - 14 = 0$

$\qquad\qquad\quad 11x = 14$

$\qquad\qquad\quad x = \dfrac{14}{11}$

17. $7x^2 - 3x = 5x$

$7x^2 - 8x = 0$

$x(7x - 8) = 0$

$x = 0 \qquad 7x - 8 = 0$

$\qquad\qquad\quad 7x = 8$

$\qquad\qquad\quad x = \dfrac{8}{7}$

19. $15x^2 + 16x = 3x^2 - 20x$

$12x^2 + 36x = 0$

$12x(x + 3) = 0$

$12x = 0 \qquad x + 3 = 0$

$x = 0 \qquad\quad x = -3$

21. $x^2 + 5x - 24 = 0$

$(x + 8)(x - 3) = 0$

$x + 8 = 0 \qquad x - 3 = 0$

$x = -8 \qquad\quad x = 3$

23. $2x^2 - 15x - 8 = 0$

$(2x + 1)(x - 8) = 0$

$2x + 1 = 0 \qquad x - 8 = 0$

$2x = -1 \qquad\quad x = 8$

$x = -\dfrac{1}{2}$

25. $3x^2 - 13x + 4 = 0$

$(3x - 1)(x - 4) = 0$

$3x - 1 = 0 \qquad x - 4 = 0$

$3x = 1 \qquad\quad x = 4$

$x = \dfrac{1}{3}$

27. $x^2 - 9x + 14 = 0$

$(x - 2)(x - 7) = 0$

$x - 2 = 0 \qquad x - 7 = 0$

$x = 2 \qquad\quad x = 7$

29. $0 = x^2 + 15x + 50$

$0 = (x + 10)(x + 5)$

$x + 10 = 0 \qquad x + 5 = 0$

$x = -10 \qquad\quad x = -5$

31. $0 = 6x^2 + 19x + 10$

$0 = (3x + 2)(2x + 5)$

$3x + 2 = 0 \qquad 2x + 5 = 0$

$3x = -2 \qquad\quad 2x = -5$

$x = -\dfrac{2}{3} \qquad x = -\dfrac{5}{2}$

33. $y^2 + 8y + 13 = y + 1$

$y^2 + 7y + 12 = 0$

$(y + 3)(y + 4) = 0$

$y + 3 = 0 \qquad y + 4 = 0$

$y = -3 \qquad\quad y = -4$

35. $\qquad 6y^2 = 11y - 3$

$6y^2 - 11y + 3 = 0$

$(3y - 1)(2y - 3) = 0$

$3y - 1 = 0 \qquad 2y - 3 = 0$

$3y = 1 \qquad\quad 2y = 3$

$y = \dfrac{1}{3} \qquad\quad y = \dfrac{3}{2}$

37. $9y^2 - 24y + 16 = 0$

$(3y - 4)(3y - 4) = 0$

$3y - 4 = 0$

$3y = 4$

$y = \dfrac{4}{3}$

39. $n^2 - 10n + 25 = 0$

$(n - 5)(n - 5) = 0$

$n - 5 = 0$

$n = 5$

41. $n^2 + 5n - 17 = 6n + 3$

$n^2 - n - 20 = 0$

$(n - 5)(n + 4) = 0$

$n - 5 = 0 \qquad n + 4 = 0$

$n = 5 \qquad\quad n = -4$

43. $3x^2 + 8x - 10 = -6x - 10$

$3x^2 + 14x = 0$

$x(3x + 14) = 0$

$x = 0 \qquad 3x + 14 = 0$

$\qquad\qquad 3x = -14$

$\qquad\qquad x = -\dfrac{14}{3}$

45. $\qquad 2x(x - 4) = 4 - x$

$2x^2 - 8x = 4 - x$

$2x^2 - 7x - 4 = 0$

$(2x + 1)(x - 4) = 0$

$2x + 1 = 0 \qquad x - 4 = 0$

$2x = -1 \qquad\quad x = 4$

$x = -\dfrac{1}{2}$

47. $(x-6)(x+1)=8$

$\quad x^2-5x-6=8$

$\quad x^2-5x-14=0$

$\quad (x-7)(x+2)=0$

$\quad x-7=0 \qquad x+2=0$

$\qquad x=7 \qquad\quad x=-2$

49. $\quad x(x+8)=16(x-1)$

$\quad\ x^2+8x=16x-16$

$\quad x^2-8x+16=0$

$\quad (x-4)(x-4)=0$

$\qquad x-4=0$

$\qquad\quad x=4$

51. $2x(x+3)=(3x+1)(x+1)$

$\quad 2x^2+6x=3x^2+4x+1$

$\qquad\ 0=x^2-2x+1$

$\qquad\ 0=(x-1)(x-1)$

$\qquad\ 0=x-1$

$\qquad\ 1=x$

53. $-3x+(x+6)(x+2)=12-3x$

$\quad -3x+x^2+8x+12=12-3x$

$\qquad\qquad\ x^2+8x=0$

$\qquad\qquad\ x(x+8)=0$

$\quad x=0 \qquad x+8=0$

$\qquad\qquad\quad x=-8$

55. $\quad \dfrac{4}{x}+\dfrac{3}{x+5}=2$

$\quad (x+5)(4)+3x=2x(x+5)$

$\quad\ 4x+20+3x=2x^2+10x$

$\quad\ 2x^2+3x-20=0$

$\quad (2x-5)(x+4)=0$

$\quad 2x-5=0 \qquad x+4=0$

$\qquad x=\dfrac{5}{2} \qquad\quad x=-4$

55. continued

Check: $\dfrac{4}{-4}+\dfrac{3}{-4+5}\overset{?}{=}2 \qquad \dfrac{4}{5/2}+\dfrac{3}{5/2+5}\overset{?}{=}2$

$\qquad\qquad\qquad 2=2 \qquad\qquad\qquad\qquad 2=2$

57. $\dfrac{12}{x-2}=\dfrac{2x-3}{3}$

$\quad 3(12)=(x-2)(2x-3)$

$\qquad\ 36=2x^2-7x+6$

$\qquad\ 0=2x^2-7x-30$

$\qquad\ 0=(2x+5)(x-6)$

$\quad 2x+5=0 \qquad x-6=0$

$\qquad x=-\dfrac{5}{2} \qquad\quad x=6$

Check: $\dfrac{12}{-5/2-2}\overset{?}{=}\dfrac{2(-5/2)-3}{3} \qquad \dfrac{12}{6-2}\overset{?}{=}\dfrac{2(6)-3}{3}$

$\qquad\qquad -\dfrac{8}{3}=-\dfrac{8}{3} \qquad\qquad\qquad 3=3$

59. $\quad \dfrac{x}{2}+\dfrac{5}{2}=-\dfrac{3}{x}$

$\quad x^2+5x=-6$

$\quad x^2+5x+6=0$

$\quad (x+3)(x+2)=0$

$\quad x+3=0 \qquad x+2=0$

$\qquad x=-3 \qquad\quad x=-2$

Check: $\dfrac{-3}{2}+\dfrac{5}{2}\overset{?}{=}\dfrac{-3}{-3} \qquad \dfrac{-2}{2}+\dfrac{5}{2}\overset{?}{=}-\dfrac{3}{-2}$

$\qquad\qquad 1=1 \qquad\qquad\qquad\qquad \dfrac{3}{2}=\dfrac{3}{2}$

61. $\quad x-8=-\dfrac{15}{x}$

$\quad x(x-8)=-15$

$\quad x^2-8x=-15$

$\quad x^2-8x+15=0$

$\quad (x-3)(x-5)=0$

61. continued

$$x - 3 = 0 \qquad x - 5 = 0$$
$$x = 3 \qquad\qquad x = 5$$

Check: $3 - 8 \overset{?}{=} -\dfrac{15}{3} \qquad 5 - 8 \overset{?}{=} -\dfrac{15}{5}$

$$-5 = -5 \qquad\qquad -3 = -3$$

63. $\dfrac{24}{x^2 - 4} = 1 + \dfrac{2x - 6}{x - 2}$

$$24 = (x^2 - 4)(1) + (x + 2)(2x - 6)$$
$$24 = x^2 - 4 + 2x^2 - 2x - 12$$
$$0 = 3x^2 - 2x - 40$$
$$0 = (3x + 10)(x - 4)$$

$$3x + 10 = 0 \qquad x - 4 = 0$$
$$x = -\dfrac{10}{3} \qquad\qquad x = 4$$

Check:

$$\dfrac{24}{(-10/3)^2 - 4} \overset{?}{=} 1 + \dfrac{2(-10/3) - 6}{-10/3 - 2} \qquad \dfrac{24}{4^2 - 4} \overset{?}{=} 1 + \dfrac{2(4) - 6}{4 - 2}$$

$$\dfrac{27}{8} = \dfrac{27}{8} \qquad\qquad\qquad 2 = 2$$

To Think About

65. You can always factor out x.

67. $ax^2 - 7x + c = 0$

$$x = -\dfrac{3}{2} \Rightarrow \dfrac{9}{4}a + \dfrac{21}{2} + c = 0 \qquad (1)$$
$$x = 6 \Rightarrow 36a - 42 + c = 0 \qquad (2)$$

Multiply (1) by -1 and add it to (2)

$$\dfrac{135}{4}a - \dfrac{105}{2} = 0$$
$$a = \dfrac{14}{9}$$

Substitute $\dfrac{14}{9}$ for a in (2)

$$36\left(\dfrac{14}{9}\right) - 42 + c = 0$$
$$c = -14$$

$$a = \dfrac{14}{9}, \quad c = -14$$

69. $t = \dfrac{n^2 - n}{2}, \quad n = 22$

$$t = \dfrac{(22)^2 - 22}{2} = 231$$

A maximum of 231 truck routes.

71. $t = \dfrac{n^2 - n}{2}, \quad t = 15$

$$15 = \dfrac{n^2 - n}{2}$$
$$0 = n^2 - n - 30$$
$$0 = (n - 6)(n + 5)$$

$$n - 6 = 0 \qquad n + 5 = 0$$
$$n = 6 \qquad\quad n = -5, \text{ has no meaning}$$

6 cities can be serviced.

73. $t = \dfrac{n^2 - n}{2}, \quad t = 55$

$$55 = \dfrac{n^2 - n}{2}$$
$$0 = n^2 - n - 110$$
$$0 = (n - 11)(n + 10)$$

$$n - 11 = 0 \qquad n + 10 = 0$$
$$n = 11 \qquad\quad n = -10, \text{ has no meaning}$$

11 cities can be serviced.

Cumulative Review Problems

75. $\dfrac{\frac{1}{2x} + \frac{3}{4x}}{\frac{7}{x} - \frac{3}{2x}} = \dfrac{\frac{1}{2x} + \frac{3}{4x}}{\frac{7}{x} - \frac{3}{2x}} \cdot \dfrac{4x}{4x}$

$$= \dfrac{2 + 3}{28 - 6}$$
$$= \dfrac{5}{22}$$

190

77. $\dfrac{2x}{3x-5} + \dfrac{2}{3x^2-11x+10}$

$= \dfrac{2x}{3x-5} \cdot \dfrac{x-2}{x-2} + \dfrac{2}{(3x-5)(x-2)}$

$= \dfrac{2x^2-4x+2}{3x^2-11x+10}$

Exercises 10.2

1. $x^2 = 49$

$x = \pm\sqrt{49}$

$x = \pm 7$

3. $x^2 = 98$

$x = \pm\sqrt{98}$

$x = \pm 7\sqrt{2}$

5. $x^2 - 75 = 0$

$x^2 = 75$

$x = \pm\sqrt{75}$

$x = \pm 5\sqrt{3}$

7. $3x^2 = 75$

$x^2 = 25$

$x = \pm\sqrt{25}$

$x = \pm 5$

9. $6x^2 = 120$

$x^2 = 20$

$x = \pm\sqrt{20}$

$x = \pm 2\sqrt{5}$

11. $4x^2 - 12 = 0$

$4x^2 = 12$

$x^2 = 3$

$x = \pm\sqrt{3}$

13. $2x^2 - 196 = 0$

$2x^2 = 196$

$x^2 = 98$

$x = \pm\sqrt{98}$

$x = \pm 7\sqrt{2}$

15. $2x^2 - 10 = 62$

$2x^2 = 72$

$x^2 = 36$

$x = \pm\sqrt{36}$

$x = \pm 6$

17. $5x^2 + 13 = 73$

$5x^2 = 60$

$x^2 = 12$

$x = \pm\sqrt{12}$

$x = \pm 2\sqrt{3}$

19. $13x^2 + 17 = 82$

$13x^2 = 65$

$x^2 = 5$

$x = \pm\sqrt{5}$

21. $(x-3)^2 = 5$

$x - 3 = \pm\sqrt{5}$

$x = 3 \pm \sqrt{5}$

23. $(x+6)^2 = 4$

$x + 6 = \pm\sqrt{4}$

$x + 6 = \pm 2$

$x = -6 \pm 2$

$x = -6 + 2 \qquad x = -6 - 2$

$x = -4 \qquad\quad x = -8$

25. $(2x+5)^2 = 2$

$$2x+5 = \pm\sqrt{2}$$
$$2x = -5 \pm \sqrt{2}$$
$$x = \frac{-5 \pm \sqrt{2}}{2}$$

27. $(3x-1)^2 = 7$

$$3x-1 = \pm\sqrt{7}$$
$$3x = 1 \pm \sqrt{7}$$
$$x = \frac{1 \pm \sqrt{7}}{3}$$

29. $(7x+2)^2 = 12$

$$7x+2 = \pm\sqrt{12}$$
$$7x = -2 \pm 2\sqrt{3}$$
$$x = \frac{-2 \pm 2\sqrt{3}}{7}$$

31. $(6x-5)^2 = 18$

$$6x-5 = \pm\sqrt{18}$$
$$6x = 5 \pm 3\sqrt{2}$$
$$x = \frac{5 \pm 3\sqrt{2}}{6}$$

33. $x^2 - 19.263 = 0$

$$x^2 = 19.263$$
$$x = \pm\sqrt{19.263}$$
$$x = \pm 4.389$$

35. $3.120x^2 - 7.986 = 0$

$$3.120x^2 = 7.986$$
$$x^2 = 2.5596$$
$$x = \pm\sqrt{2.5596}$$
$$x = \pm 1.600$$

37. $x^2 + 14x = 15, \quad \left[\frac{1}{2}(14)\right]^2 = 7^2 = 49$

$$x^2 + 14x + 49 = 15 + 49$$
$$(x+7)^2 = 64$$
$$x+7 = \pm\sqrt{64}$$
$$x+7 = \pm 8$$
$$x = -7 \pm 8$$

$$x = -7+8 \qquad x = -7-8$$
$$x = 1 \qquad\quad x = -15$$

39. $x^2 - 4x = 1, \quad \left[\frac{1}{2}(4)\right]^2 = 2^2 = 4$

$$x^2 - 4x + 4 = 1 + 4$$
$$(x-2)^2 = 5$$
$$x-2 = \pm\sqrt{5}$$
$$x = 2 \pm \sqrt{5}$$

41. $x^2 + 6x + 7 = 0$

$$x^2 + 6x = -7, \quad \left[\frac{1}{2}(6)\right]^2 = 3^2 = 9$$
$$x^2 + 6x + 9 = -7 + 9$$
$$(x+3)^2 = 2$$
$$x+3 = \pm\sqrt{2}$$
$$x = -3 \pm \sqrt{2}$$

43. $x^2 - 10x - 4 = 0$

$$x^2 - 10x = 4, \quad \left[\frac{1}{2}(10)\right]^2 = 5^2 = 25$$
$$x^2 - 10x + 25 = 4 + 25$$
$$(x-5)^2 = 29$$
$$x-5 = \pm\sqrt{29}$$
$$x = 5 \pm \sqrt{29}$$

45. $x^2 + 3x = 0, \left[\frac{1}{2}(3)\right]^2 = \left(\frac{3}{2}\right)^2 = \frac{9}{4}$

$x^2 + 3x + \frac{9}{4} = 0 + \frac{9}{4}$

$\left(x + \frac{3}{2}\right)^2 = \frac{9}{4}$

$x + \frac{3}{2} = \pm\sqrt{\frac{9}{4}}$

$x + \frac{3}{2} = \pm\frac{3}{2}$

$x = -\frac{3}{2} \pm \frac{3}{2}$

$x = -\frac{3}{2}$ $x = -\frac{3}{2} - \frac{3}{2}$

$x = 0$ $x = -\frac{6}{2}$

           $x = -3$

47. $x^2 - x - 6 = 0$

$x^2 - x = 6, \left[\frac{1}{2}(1)\right]^2 = \left(\frac{1}{2}\right)^2 = \frac{1}{4}$

$x^2 - x + \frac{1}{4} = 6 + \frac{1}{4}$

$\left(x - \frac{1}{2}\right)^2 = \frac{25}{4}$

$x - \frac{1}{2} = \pm\sqrt{\frac{25}{4}}$

$x = \frac{1}{2} \pm \frac{5}{2}$

$x = \frac{1}{2} + \frac{5}{2}$

$x = \frac{1}{2} + \frac{5}{2}$ $x = \frac{1}{2} - \frac{5}{2}$

$x = 3$ $x = -2$

49. $4x^2 - 8x + 3 = 0$

$4x^2 - 8x = -3$

$x^2 - 2x = -\frac{3}{4}, \left[\frac{1}{2}(2)\right]^2 = 1^2 = 1$

$x^2 - 2x + 1 = -\frac{3}{4} + 1$

$(x - 1)^2 = \frac{1}{4}$

$x - 1 = \pm\sqrt{\frac{1}{4}}$

$x = 1 \pm \frac{1}{2}$

$x = 1 + \frac{1}{2}$ $x = 1 - \frac{1}{2}$

$x = \frac{3}{2}$ $x = \frac{1}{2}$

51. $2x^2 + 10x + 11 = 0$

$2x^2 + 10x = -11$

$x^2 + 5x = -\frac{11}{2}, \left[\frac{1}{2}(5)\right]^2 = \frac{25}{4}$

$x^2 + 5x + \frac{25}{4} = -\frac{11}{2} + \frac{25}{4}$

$\left(x + \frac{5x}{2}\right)^2 = \frac{3}{4}$

$x + \frac{5}{2} = \pm\sqrt{\frac{3}{4}}$

$x = -\frac{5}{2} \pm \frac{\sqrt{3}}{2}$

$x = \frac{-5 \pm \sqrt{3}}{2}$

To Think About

53. $x^2 + bx - 7 = 0$

$$x^2 + bx = 7, \quad \left(\frac{b}{2}\right)^2 = \frac{b^2}{4}$$

$$x^2 + bx + \frac{b^2}{4} = 7 + \frac{b^2}{4}$$

$$\left(x + \frac{b}{2}\right)^2 = \frac{28 + b^2}{4}$$

$$x + \frac{b}{2} = \pm\sqrt{\frac{28 + b^2}{4}}$$

$$x = -\frac{b}{2} \pm \frac{\sqrt{b^2 + 28}}{2}$$

$$x = \frac{-b \pm \sqrt{b^2 + 28}}{2}$$

Cumulative Review Problems

55. $3a - 5b = 8 \qquad (1)$

$5a - 7b = 8 \qquad (2)$

Multiply (1) by 5 and (2) by -3

$15a - 25b = 40$

$\underline{-15a + 21b = -24}$

$\qquad -4b = 16$

$\qquad b = -4$

Substitute -4 for b in (1)

$3a - 5(-4) = 8$

$\qquad 3a = -12$

$\qquad a = -4$

$a = -4, \quad b = -4$

57. $\left(\dfrac{3}{4}\right)^{-2} = \left(\dfrac{4}{3}\right)^2 = \dfrac{4^2}{3^2} = \dfrac{16}{9}$

Exercises 10.3

Use $x = \dfrac{-b \pm \sqrt{b^2 - 4ac}}{2a}$ in Exercises 1-31.

1. $x^2 + 4x + 1 = 0$

$a = 1, \quad b = 4, \quad c = 1$

$$x = \frac{-4 \pm \sqrt{16 - 4(1)(1)}}{2(1)} = \frac{-4 \pm \sqrt{12}}{2}$$

$$= \frac{-4 \pm 2\sqrt{3}}{2} = -2 \pm \sqrt{3}$$

3. $x^2 - 5x - 2 = 0$

$a = 1, \quad b = -5, \quad c = -2$

$$x = \frac{5 \pm \sqrt{25 - 4(1)(-2)}}{2(1)} = \frac{5 \pm \sqrt{33}}{2}$$

5. $2x^2 - 7x - 9 = 0$

$a = 2, \quad b = -7, \quad c = -9$

$$x = \frac{7 \pm \sqrt{49 - 4(2)(-9)}}{2(2)} = \frac{7 \pm \sqrt{121}}{4} = \frac{7 \pm 11}{4}$$

$$x = \frac{7 + 11}{4} \qquad x = \frac{7 - 11}{4}$$

$$x = \frac{9}{2} \qquad\qquad x = -1$$

7. $5x^2 + x - 1 = 0$

$a = 5, \quad b = 1, \quad c = -1$

$$x = \frac{-1 \pm \sqrt{1 - 4(5)(-1)}}{2(5)} = \frac{-1 \pm \sqrt{21}}{10}$$

9.
$$5x - 1 = 2x^2$$
$$-2x^2 + 5x - 1 = 0$$
$$a = -2, \ b = 5 \ c = -1$$
$$x = \frac{-5 \pm \sqrt{25 - 4(-2)(-1)}}{2(-2)} = \frac{-5 \pm \sqrt{17}}{-4}$$
$$= \frac{-5 \pm \sqrt{17}}{-4} \cdot \frac{(-1)}{(-1)} = \frac{5 \pm \sqrt{17}}{4}$$

11.
$$6x^2 - 3x = 1$$
$$6x^2 - 3x - 1 = 0$$
$$a = 6, \ b = -3, \ c = -1$$
$$x = \frac{3 \pm \sqrt{9 - 4(6)(-1)}}{2(6)} = \frac{3 \pm \sqrt{33}}{12}$$

13.
$$x - 1 = -\frac{2x^2}{5}$$
$$\frac{2x^2}{5} + x - 1 = 0$$
$$2x^2 + 5x - 5 = 0$$
$$a = 2, \ b = 5, \ c = -5$$
$$x = \frac{-5 \pm \sqrt{25 - 4(2)(-5)}}{2(2)} = \frac{-5 \pm \sqrt{65}}{4}$$

15.
$$9x^2 - 6x = 2$$
$$9x^2 - 6x - 2 = 0$$
$$a = 9, \ b = -6, \ c = -2$$
$$x = \frac{6 \pm \sqrt{36 - 4(9)(-2)}}{2(9)} = \frac{6 \pm \sqrt{108}}{18}$$
$$= \frac{6 \pm 6\sqrt{3}}{18} = \frac{1 \pm \sqrt{3}}{3}$$

17.
$$4x^2 + 3x + 2 = 0$$
$$a = 4, \ b = 3, \ c = 2$$
$$x = \frac{-3 \pm \sqrt{9 - 4(4)(2)}}{2(4)} = \frac{-3 \pm \sqrt{-23}}{8}$$
No real solution.

19.
$$\frac{x}{3} + \frac{2}{x} = \frac{7}{3}$$
$$x^2 + 6 = 7x$$
$$x^2 - 7x + 6 = 0$$
$$a = 1, \ b = -7, \ c = 6$$
$$x = \frac{7 \pm \sqrt{49 - 4(1)(6)}}{2(1)} = \frac{7 \pm \sqrt{25}}{2} = \frac{7 \pm 5}{2}$$
$$x = \frac{7 + 5}{2} \qquad x = \frac{7 - 5}{2}$$
$$x = 6 \qquad\qquad x = 1$$

21.
$$5y^2 = 3 - 7y$$
$$5y^2 + 7y - 3 = 0$$
$$a = 5, \ b = 7, \ c = -3$$
$$y = \frac{-7 \pm \sqrt{49 - 4(5)(-3)}}{2(5)} = \frac{-7 \pm \sqrt{109}}{10}$$

23.
$$\frac{1}{3}m^2 = \frac{1}{2}m + \frac{3}{2}$$
$$2m^2 = 3m + 9$$
$$2m^2 - 3m - 9 = 0$$
$$a = 2, \ b = -3, \ c = -9$$
$$m = \frac{3 \pm \sqrt{9 - 4(2)(-9)}}{2(2)} = \frac{3 \pm \sqrt{81}}{4} = \frac{3 \pm 9}{4}$$
$$m = \frac{3 + 9}{4} = 3 \qquad m = \frac{3 - 9}{4}$$
$$m = 3 \qquad\qquad m = -\frac{3}{2}$$

25. $1.04x^2 - 17.982x + 74.108 = 0$

$a = 1.04, \ b = -17.982, \ c = 74.108$

$$x = \frac{-(17.982) \pm \sqrt{(-17.982)^2 - 4(1.04)(74.108)}}{2(1.04)}$$

$$= \frac{17.982 \pm \sqrt{15.0630}}{2.08}$$

$x = 10.511, \ 6.779$

27. $x^2 + 3x - 7 = 0$

$a = 1, \ b = 3, \ c = -7$

$$x = \frac{-3 \pm \sqrt{9 - 4(1)(-7)}}{2(1)} = \frac{-3 \pm \sqrt{37}}{2}$$

$x = 1.541, \ -4.541$

29. $4x^2 - 4x - 1 = 0$

$a = 4, \ b = -4, \ c = -1$

$$x = \frac{4 \pm \sqrt{16 - 4(4)(-1)}}{2(4)} = \frac{4 \pm \sqrt{32}}{8}$$

$$= \frac{4 \pm 4\sqrt{2}}{8} = \frac{1 \pm \sqrt{2}}{2}$$

$x = 1.207, \ -0.207$

31. $5x^2 + 10x + 1 = 0$

$a = 5, \ b = 10, \ c = 1$

$$x = \frac{-10 \pm \sqrt{100 - 4(5)(1)}}{2(5)} = \frac{-10 \pm \sqrt{80}}{10}$$

$$= \frac{-10 \pm 4\sqrt{5}}{10} = \frac{-5 \pm 2\sqrt{5}}{5}$$

$x = -0.106, \ -1.894$

33. $$t^2 + 1 = \frac{13t}{6}$$

$$6t^2 + 6 = 13t$$

$$6t^2 - 13t + 6 = 0$$

$$(3t - 2)(2t - 3) = 0$$

$3t - 2 = 0 \qquad 2t - 3 = 0$

$t = \dfrac{2}{3} \qquad\qquad t = \dfrac{3}{2}$

35. $$3(s^2 + 1) = 10s$$

$$3s^2 + 3 = 10s$$

$$3s^2 - 10s + 3 = 0$$

$$(3s - 1)(s - 3) = 0$$

$3s - 1 = 0 \qquad s - 3 = 0$

$s = \dfrac{1}{3} \qquad\qquad s = 3$

37. $(p - 2)(p + 1) = 3$

$$p^2 - p - 2 = 3$$

$$p^2 - p - 5 = 0$$

$a = 1, \ b = -1, \ c = -5$

$$p = \frac{1 \pm \sqrt{1 - 4(1)(-5)}}{2(1)} = \frac{1 \pm \sqrt{21}}{2}$$

39. $$\frac{7y^2}{3} - \frac{8}{3} = y$$

$$7y^2 - 8 = 3y$$

$$7y^2 - 3y - 8 = 0$$

$a = 7, \ b = -3, \ c = -8$

$$y = \frac{3 \pm \sqrt{9 - 4(7)(-8)}}{2(7)} = \frac{3 \pm \sqrt{233}}{14}$$

41. $3x^2 - 7 = 0$

$a = 3, \ b = 0, \ c = -7$

$$x = \frac{0 \pm \sqrt{0 - 4(3)(-7)}}{2(3)} = \frac{\pm\sqrt{84}}{6}$$

$$= \frac{\pm 2\sqrt{21}}{6} = \frac{\pm\sqrt{21}}{3}$$

43.
$$x(x-2) = 7$$
$$x^2 - 2x = 7$$
$$x^2 - 2x - 7 = 0$$
$$a = 1, \ b = -2, \ c = -7$$
$$x = \frac{-(-2) \pm \sqrt{4 - 4(1)(-7)}}{2(1)}$$
$$= \frac{2 \pm \sqrt{32}}{2} = \frac{2 \pm 4\sqrt{2}}{2}$$
$$= 1 \pm 2\sqrt{2}$$

Cumulative Review Problems

45.
$$\frac{3x}{x-2} + \frac{4}{x+2} - \frac{x+22}{x^2-4}$$
$$= \frac{3x}{x-2} \cdot \frac{x+2}{x+2} + \frac{4}{x+2} \cdot \frac{x-2}{x-2} - \frac{x+22}{x^2-4}$$
$$= \frac{3x^2 + 6x + 4x - 8 - x - 22}{(x-2)(x+2)}$$
$$= \frac{3x^2 + 9x - 30}{(x-2)(x+2)}$$
$$= \frac{3(x+5)(x-2)}{(x-2)(x+2)}$$
$$= \frac{3(x+5)}{x+2}$$

47.
$$\begin{array}{r} x^2 + 5x + 2 \\ x+3{\overline{\smash{\big)}\,x^3 + 8x^2 + 17x + 6}} \\ \underline{x^3 + 3x^2} \\ 5x^2 + 17x \\ \underline{5x^2 + 15x} \\ 2x + 6 \\ \underline{2x + 6} \end{array}$$

$$\left(x^3 + 8x^2 + 17x + 6\right) \div (x+3) = x^2 + 5x + 2$$

Exercises 10.4

1. $y = 2x^2 - 1$

x	y
-2	7
-1	1
0	-1
1	1
2	7

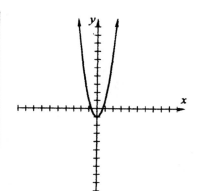

3. $y = -\frac{1}{3}x^2$

x	y
-6	-12
-3	-3
0	0
3	-3
6	-12

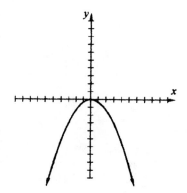

5. $y = x^2 - 2$

x	y
-3	7
-1	-1
0	-2
1	-1
3	7

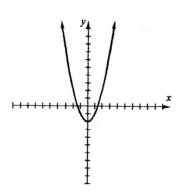

7. $y = (x-2)^2$

197

x	y
0	4
1	1
2	0
3	1
4	4

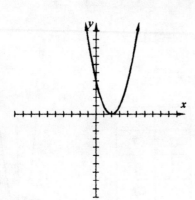

13. $y = x^2 + 4x, \quad a = 1, \quad b = 4$

$a > 0$: opens up

Vertex: $x = -\dfrac{b}{2a} = -\dfrac{4}{2(1)} = -2$

$\qquad y = (-2)^2 + 4(-2) = -4$

$V(-2, -4)$

x	y
-4	0
-3	-3
-2	-4
-1	-3
0	0

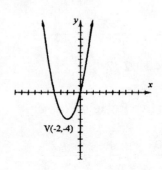

9. $y = -\dfrac{1}{2}x^2 + 4$

x	y
-4	-4
-2	2
0	4
2	2
4	-4

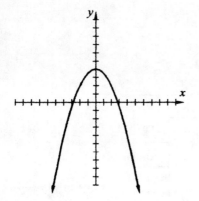

15. $y = x^2 - 4x, \quad a = 1, \quad b = -4$

$a > 0$: opens up

Vertex: $x = -\dfrac{(-4)}{2(1)} = 2$

$\qquad y = (2)^2 - 4(2) = -4$

$V(2, -4)$

x	y
0	0
1	-3
2	-4
3	-3
4	0

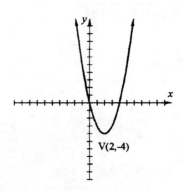

11. $y = \dfrac{1}{2}(x - 3)^2$

x	y
-1	8
1	2
3	0
5	2
7	8

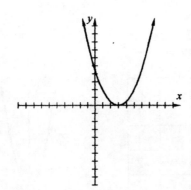

198

17. $y = -x^2 + 6x, \quad a = -1, \quad b = 6$

$a < 0$: opens down

Vertex: $x = -\dfrac{6}{2(-1)} = 3$

$\qquad y = -3^2 + 6(3) = 9$

$V(3, 9)$

x	y
0	0
1	5
3	9
5	5
6	0

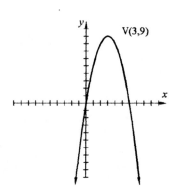

19. $y = x^2 - 4x + 3, \quad a = 1, \quad b = -4$

Vertex: $x = -\dfrac{-4}{2(1)} = 2$

$\qquad y = 2^2 - 4(2) + 3 = -1$

$V(2, -1)$

$y - $ intercept: $y = 0^2 - 4(0) + 3 = 3$

$x - $ intercepts: $0 = x^2 - 4x + 3$

$\qquad\qquad 0 = (x - 3)(x - 1)$

$\qquad\qquad x = 1, 3$

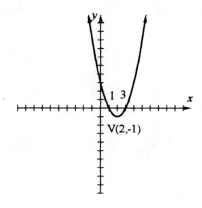

21. $y = -x^2 - 6x - 5, \quad a = -1, \quad b = -6$

Vertex: $x = -\dfrac{(-6)}{2(-1)} = -3$

$\qquad y = -(-3)^2 - 6(-3) - 5 = 4$

$V(-3, 4)$

$y - $ intercept: $y = -0^2 - 6(0) - 5 = -5$

$x - $ intercepts: $0 = -x^2 - 6x - 5$

$\qquad\qquad 0 = -(x + 5)(x + 1)$

$\qquad\qquad x = -5, -1$

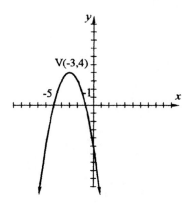

23. $y = -x^2 + 4x - 4, \quad a = -1, \quad b = 4$

Vertex: $x = -\dfrac{4}{2(-1)} = 2$

$\qquad y = -2^2 + 4(2) - 4 = 0$

$V(2, 0)$

$y - $ intercept: $y = -0^2 + 4(0) - 4 = -4$

$x - $ intercepts: $0 = -x^2 + 4x - 4$

$\qquad\qquad 0 = -(x - 2)^2$

$\qquad\qquad x = 2$

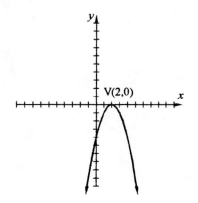

25. $h = -4.9t^2 + 19.6t + 10,\quad a = -4.9,\quad b = 19.6$

(a) Vertex: $t = -\dfrac{19.6}{2(-4.9)} = 2$

$$h = -4.9(2)^2 + 19.6(2) + 10 = 29.6$$

$V(2,\ 29.6)$

h – intercept: $h = 10$

t – intercept: $0 = -4.9t^2 + 19.6t + 10$

$$t \approx -0.5,\ 4.5$$

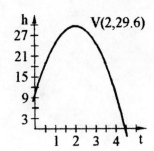

(b) $t = 3$

$$h = -4.9(3)^2 + 19.6(3) + 10 = 24.7$$

It is 24.7 m high.

(c) $V(2,\ 29.6)$

The highest point is 29.6 m.

(d) The x – intercept is 4.5

After $4.5s$ it will strike the earth.

To Think About

27. $y = -2x^2 + 4x + 5,\quad a = -2,\quad b = 4$

Vertex: $x = -\dfrac{4}{2(-2)} = 1$

$$y = -2(1)^2 + 4(1) + 5 = 7$$

$V(1,\ 7)$

x – intercepts: $0 = -2x^2 + 4x + 5$

$$x = \dfrac{-4 \pm \sqrt{16 - 4(-2)(5)}}{2(-2)}$$

$$= \dfrac{-4 \pm \sqrt{56}}{-4}$$

$$= \dfrac{2 \pm \sqrt{14}}{2}$$

Cumulative Review Problems

29. $\sqrt{2x+3} + 3 = 7$

$$\sqrt{2x+3} = 4$$

$$2x + 3 = 16$$

$$2x = 13$$

$$x = \dfrac{13}{2}$$

Check: $\sqrt{2\left(\dfrac{13}{2}\right) + 3} + 3 \overset{?}{=} 7$

$$7 = 7$$

31. $y = \dfrac{k}{x},\quad y = \dfrac{1}{3},\quad x = 21$

$$\dfrac{1}{3} = \dfrac{k}{21} \Rightarrow k = 7$$

$$y = \dfrac{7}{x},\quad x = 7$$

$$y = \dfrac{7}{7} = 1$$

200

Putting Your Skills To Work

1. $x = 8$ yd, Height is 11.5 feet

3. $h = 8$ feet, distance is 3 yards.

$$h = -\frac{1}{35}(x-15)^2 + 12,$$

$$h = -\frac{1}{35}x^2 + \frac{6}{7}x + \frac{39}{7}, \quad a = -\frac{1}{35}, \quad b = \frac{6}{7}$$

Vertex: $x = -\dfrac{6/7}{2(-1/35)} = 15$

$$h = -\frac{1}{35}(15-15)^2 + 12 = 12$$

h – intercept: $h = -\dfrac{1}{35}(0-15)^2 + 12 = \dfrac{39}{7}$

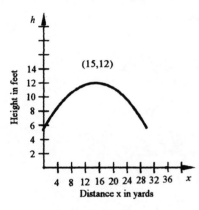

5. Maximum height is at vertex: 12 feet.

7. $x = 7$ feet, distance is 2 yards.

Exercises 10.5

1. $x = $ length on one leg
$x + 8 = $ length of other leg
$$(x+8)^2 + x^2 = 40^2$$
$$x^2 + 16x + 64 + x^2 = 1600$$
$$2x^2 + 16x - 1536 = 0$$
$$x^2 + 8x - 768 = 0$$
$$(x+32)(x-24) = 0$$

1. continued

$x = -32, \qquad x = 24$

$x + 8 = 32$

Length can't be negative.

The lengths are 24 yards, 32 yards.

3. $\qquad x = $ length

$x - 7 = $ width

$$(x-7)^2 + x^2 = 13^2$$
$$x^2 - 14x + 49 + x^2 = 169$$
$$2x^2 - 14x - 120 = 0$$
$$x^2 - 7x - 60 = 0$$
$$(x-12)(x+5) = 0$$
$$x = 12, \quad x = -5,$$

$x - 7 = 5$

Length can't be negative.

The dimensions are 12 m by 5 m.

5. $\quad s = $ number of students

$c = $ cost per students

$$s \cdot c = 420 \qquad (1)$$
$$(s+7)(c-5) = 420 \qquad (2)$$

Substitute $\dfrac{420}{5}$ for c in (2)

$$(s+7)\left(\frac{420}{s} - 5\right) = 420$$
$$(s+7)(420 - 5s) = 420s$$
$$(s+7)(84 - s) = 84s$$
$$s^2 + 7s - 588 = 0$$
$$(s+28)(s-21) = 0$$

$s = -28, \quad s = 21$

Number of students can't be negative.

Original number of students was 21.

201

7. s = number of students

c = cost of fees

$$s \cdot c = 18{,}000 \qquad (1)$$

$$(s+90)(c-10) = 18{,}000 \qquad (2)$$

Substitute $\dfrac{18{,}000}{s}$ for c in (2)

$$(s+90)\left(\frac{18{,}000}{s} - 10\right) = 18{,}000$$

$$(s+90)(18{,}000 - 10s) = 18{,}000s$$

$$(s+90)(1800 - s) = 1800s$$

$$s^2 + 90s - 162{,}000 = 0$$

$$(s+450)(s-360) = 0$$

$$s = -450, \quad s = 360$$

Number of students can't be negative.

$$s + 90 = 450$$

360 were planned for, 450 came.

9. x = width

y = length

Length: $2x + y = 140 \qquad (1)$

Area: $xy = 2000 \qquad (2)$

Substitute $\dfrac{2000}{y}$ for x in (1)

$$2\left(\frac{2000}{y}\right) + y = 140$$

$$4000 + y^2 = 140y$$

$$y^2 - 140y + 4000 = 0$$

$$(y-100)(y-40) = 0$$

$$y = 100 \qquad\qquad y = 40$$

$$x = \frac{2000}{y} = 20 \qquad x = \frac{2000}{y} = 50$$

If length is 100 feet, width is 20 feet.

If length is 40 feet, width is 50 feet.

11. x = actual speed

$x + 20$ = proposed speed

Actual test: time $= \dfrac{2400}{x}$

Proposed test: time $= \dfrac{2400}{x+200}$

$$\frac{2400}{x+200} = \frac{2400}{x} - 1$$

$$2400x = 2400(x+200) - x(x+200)$$

$$2400x = 2400x + 480{,}000 - x^2 - 200x$$

$$0 = -x^2 - 200x + 480{,}000$$

$$0 = -(x+800)(x-600)$$

$$x = -800 \qquad x = 600$$

Speed can't be negative.

The actual speed was 600 mph.

13. $\dfrac{n(n+1)}{2} = s, \quad s = 91$

$$\frac{n(n+1)}{2} = 91$$

$$n^2 + n = 182$$

$$n^2 + n - 182 = 0$$

$$(n+14)(x-13) = 0$$

$$n = -14 \qquad n = 13$$

Counting numbers can't be negative.

$$n = 13$$

15. $p = -2x^2 + 360x - 14{,}400, \quad x = 85$

$$p = -2(85)^2 + 360(85) - 14{,}400$$

$$p = 1750$$

Profit is \$1750.

17. $3x + 11 \geq 9x - 4$

$-6x \geq -15$

$x \leq \dfrac{5}{2}$

$$0 \qquad \tfrac{5}{2}$$

Putting Your Skills To Work

1. $t = $ height

$\dfrac{1.618}{1} = \dfrac{26}{t}$

$1.618t = 26$

$t = 16.069$

It should be 16.069 inches tall.

3. $l = $ length

$w = $ width

$\dfrac{1.618}{1} = \dfrac{l}{w} \Rightarrow l = 1.618$ \qquad (1)

$l^2 + w^2 = (21)^2 \Rightarrow l^2 + w^2 = 441$ \quad (2)

Substitute $1.618w$ for l in (2)

$(1.618w)^2 + w^2 = 441$

$2.618w^2 + w^2 = 441$

$3.618w^2 = 441$

$w^2 = 121.89$

$w = \pm 11.04$

width can't be negative.

$w = 11.04$

$l = 1.618w = 17.86$

width $= 11.04$

length $= 17.86$

1. \qquad $5x^2 = -2x + 3$

$5x^2 + 2x - 3 = 0$

$a = 5, \ b = 2, \ c = -3$

3. $(x - 4)(2x - 1) = x^2 + 7$

$2x^2 - 9x + 4 = x^2 + 7$

$x^2 - 9x - 3 = 0$

$a = 1, \ b = -9, \ c = -3$

5. \qquad $\dfrac{1}{x^2} - \dfrac{3}{x} + 5 = 0$

$1 - 3x + 5x^2 = 0$

$5x^2 - 3x + 1 = 0$

$a = 5, \ b = -3, \ c = 1$

7. \qquad $x^2 + 7x + 8 = 4x^2 + x + 8$

$0 = 3x^2 - 6x$

$3x^2 - 6x = 0$

$a = 3, \ b = -6, \ c = 0$

9. $x^2 + 21x + 20 = 0$

$(x + 20)(x + 1) = 0$

$x + 20 = 0 \qquad x + 1 = 0$

$x = -20 \qquad x = -1$

11. $x^2 + 6x - 27 = 0$

$(x - 3)(x + 9) = 0$

$x - 3 = 0 \qquad x + 9 = 0$

$x = 3 \qquad x = -9$

13. \qquad $6x^2 + 23x = -15$

$6x^2 + 23x + 15 = 0$

$(6x + 5)(x + 3) = 0$

$6x + 5 = 0 \qquad x + 3 = 0$

$6x = -5 \qquad x = -3$

$x = -\dfrac{5}{6}$

15. $$12x^2 = 4 - 13x$$
$$12x^2 + 13x - 4 = 0$$
$$(3x + 4)(4x - 1) = 0$$

$$3x + 4 = 0 \qquad 4x - 1 = 0$$
$$3x = -4 \qquad 4x = 1$$
$$x = -\frac{4}{3} \qquad x = \frac{1}{4}$$

17. $$x^2 + \frac{5}{6}x - 1 = 0$$
$$6x^2 + 5x - 6 = 0$$
$$(3x - 2)(2x + 3) = 0$$

$$3x - 2 = 0 \qquad 2x + 3 = 0$$
$$3x = 2 \qquad 2x = -3$$
$$x = \frac{2}{3} \qquad x = -\frac{3}{2}$$

19. $$x^2 + \frac{2}{15}x = \frac{1}{15}$$
$$15x^2 + 2x = 1$$
$$15x^2 + 2x - 1 = 0$$
$$(5x - 1)(3x + 1) = 0$$

$$5x - 1 = 0 \qquad 3x + 1 = 0$$
$$5x = 1 \qquad 3x = -1$$
$$x = \frac{1}{5} \qquad x = -\frac{1}{3}$$

21. $$5x^2 + 16x = 0$$
$$x(5x + 16) = 0$$
$$x = 0 \qquad 5x + 16 = 0$$
$$5x = -16$$
$$x = -\frac{16}{5}$$

23. $$x^2 + 12x - 3 = 8x - 3$$
$$x^2 + 4x = 0$$
$$x(x + 4) = 0$$

$$x = 0 \qquad x + 4 = 0$$
$$x = -4$$

25. $$1 + \frac{2}{3x + 4} = \frac{3}{3x + 2}$$
$$(3x + 4)(3x + 2) + 2(3x + 2) = 3(3x + 4)$$
$$9x^2 + 18x + 8 + 6x + 4 = 9x + 12$$
$$9x^2 + 15x = 0$$
$$3x(3x + 5) = 0$$

$$3x = 0 \qquad 3x + 5 = 0$$
$$x = 0 \qquad 3x = -5$$
$$x = -\frac{5}{3}$$

27. $$5 + \frac{24}{2 - x} = \frac{24}{2 + x}$$
$$5(2 - x)(2 + x) + 24(2 + x) = 24(2 - x)$$
$$20 - 5x^2 + 48 + 24x = 48 - 24x$$
$$-5x^2 + 48x + 20 = 0$$
$$5x^2 - 48x - 20 = 0$$
$$(5x + 2)(x - 10) = 0$$

$$5x + 2 = 0 \qquad x - 10 = 0$$
$$5x = -2 \qquad x = 10$$
$$x = -\frac{2}{5}$$

29. $$3 = \frac{10}{x} + \frac{8}{x^2}$$
$$3x^2 = 10x + 8$$
$$3x^2 - 10x - 8 = 0$$
$$(3x + 2)(x - 4) = 0$$

$$3x + 2 = 0 \qquad x - 4 = 0$$
$$3x = -2 \qquad x = 4$$
$$x = -\frac{2}{3}$$

31. $$x^2 = 100$$
$$x = \pm\sqrt{100}$$
$$x = \pm 10$$

33. $x^2 + 3 = 28$

$\qquad x^2 = 25$

$\qquad x = \pm\sqrt{25}$

$\qquad x = \pm 5$

35. $x^2 - 5 = 17$

$\qquad x^2 = 22$

$\qquad x = \pm\sqrt{22}$

37. $2x^2 - 1 = 15$

$\qquad 2x^2 = 16$

$\qquad x^2 = 8$

$\qquad x = \pm\sqrt{8}$

$\qquad x = \pm 2\sqrt{2}$

39. $3x^2 + 6 = 60$

$\qquad 3x^2 = 54$

$\qquad x^2 = 18$

$\qquad x = \pm\sqrt{18}$

$\qquad x = \pm 3\sqrt{2}$

41. $(x - 4)^2 = 7$

$\qquad x - 4 = \pm\sqrt{7}$

$\qquad x = 4 \pm \sqrt{7}$

43. $(5x + 6)^2 = 20$

$\qquad 5x + 6 = \pm\sqrt{20}$

$\qquad 5x = -6 \pm 2\sqrt{5}$

$\qquad x = \dfrac{-6 \pm 2\sqrt{5}}{5}$

45. $x^2 + 6x + 5 = 0$

$\qquad x^2 + 6x = -5, \quad \left[\dfrac{1}{2}(6)\right]^2 = 3^2 = 9$

$\qquad x^2 + 6x + 9 = -5 + 9$

$\qquad (x + 3)^2 = 4$

$\qquad x + 3 = \pm\sqrt{4}$

$\qquad x = -3 \pm 2$

45. continued

$\qquad x = -3 + 2 \qquad x = -3 - 2$

$\qquad x = -1 \qquad\qquad x = -5$

47. $x^2 - 4x + 2 = 0$

$\qquad x^2 - 4x = -2, \quad \left[\dfrac{1}{2}(-4)\right]^2 = (-2)^2 = 4$

$\qquad x^2 - 4x + 4 = -2 + 4$

$\qquad (x - 2)^2 = 2$

$\qquad x - 2 = \pm\sqrt{2}$

$\qquad x = 2 \pm \sqrt{2}$

49. $2x^2 - 8x - 90 = 0$

$\qquad 2x^2 - 8x = 90$

$\qquad x^2 - 4x = 45, \quad \left[\dfrac{1}{2}(-4)\right]^2 = (-2)^2 = 4$

$\qquad x^2 - 4x + 4 = 45 + 4$

$\qquad (x - 2)^2 = 49$

$\qquad x - 2 = \pm\sqrt{49}$

$\qquad x = 2 \pm 7$

$\qquad x = 2 + 7 \qquad x = 2 - 7$

$\qquad x = 9 \qquad\qquad x = -5$

51. $3x^2 + 6x - 6 = 0$

$\qquad 3x^2 + 6x = 6$

$\qquad x^2 + 2x = 2, \quad \left[\dfrac{1}{2}(2)\right]^2 = 1^2 = 1$

$\qquad x^2 + 2x + 1 = 2 + 1$

$\qquad (x + 1)^2 = 3$

$\qquad x + 1 = \pm\sqrt{3}$

$\qquad x = -1 \pm \sqrt{3}$

Use $x = \dfrac{-b \pm \sqrt{b^2 - 4ac}}{2a}$ in Exercises 53 - 63.

53. $x^2 + 3x - 40 = 0$

$a = 1, \;\; b = 3, \;\; c = -40$

$x = \dfrac{-3 \pm \sqrt{9 - 4(1)(-40)}}{2(1)}$

$= \dfrac{-3 \pm \sqrt{169}}{2}$

$= \dfrac{-3 \pm 13}{2}$

$x = \dfrac{-3 + 13}{2} \qquad x = \dfrac{-3 - 13}{2}$

$x = 5 \qquad\qquad x = -8$

55. $x^2 + 4x - 6 = 0$

$a = 1, \;\; b = 4, \;\; c = -6$

$x = \dfrac{-4 \pm \sqrt{16 - 4(1)(-6)}}{2(1)}$

$= \dfrac{-4 \pm \sqrt{40}}{2}$

$= \dfrac{-4 \pm 2\sqrt{10}}{2}$

$x = -2 \pm \sqrt{10}$

57. $2x^2 - 7x + 4 = 0$

$a = 2, \;\; b = -7, \;\; c = 4$

$x = \dfrac{-(-7) \pm \sqrt{49 - 4(2)(4)}}{2(2)}$

$x = \dfrac{7 \pm \sqrt{17}}{4}$

59. $2x^2 + 4x - 5 = 0$

$a = 2, \;\; b = 4, \;\; c = -5$

$x = \dfrac{-4 \pm \sqrt{16 - 4(2)(-5)}}{2(2)}$

$= \dfrac{-4 \pm \sqrt{56}}{4}$

$= \dfrac{-4 \pm 2\sqrt{14}}{4}$

$x = \dfrac{-2 \pm \sqrt{14}}{2}$

61. $\quad 3x^2 - 5x = 4$

$3x^2 - 5x - 4 = 0$

$a = 3, \;\; b = -5, \;\; c = -4$

$x = \dfrac{-(-5) \pm \sqrt{25 - 4(3)(-4)}}{2(3)}$

$= \dfrac{5 \pm \sqrt{73}}{6}$

63. $\qquad 4x^2 = 1 - 4x$

$4x^2 + 4x - 1 = 0$

$a = 4, \;\; b = 4, \;\; c = -1$

$x = \dfrac{-4 \pm \sqrt{16 - 4(4)(-1)}}{2(4)}$

$= \dfrac{-4 \pm \sqrt{32}}{8}$

$= \dfrac{-4 \pm 4\sqrt{2}}{8}$

$= \dfrac{-1 \pm \sqrt{2}}{2}$

65. $\quad 2x^2 - 9x + 10 = 0$

$(2x - 5)(x - 2) = 0$

$2x - 5 = 0 \qquad x - 2 = 0$

$x = \dfrac{5}{2} \qquad\qquad x = 2$

67. $9x^2 - 6x + 1 = 0$

$(3x - 1)^2 = 0$

$3x - 1 = 0$

$x = \dfrac{1}{3}$

69. $3x^2 - 6x + 2 = 0$

$x = \dfrac{-(-6) \pm \sqrt{(-6)^2 - 4(3)(2)}}{2(3)}$

$= \dfrac{6 \pm \sqrt{12}}{6}$

$= \dfrac{6 \pm 2\sqrt{3}}{6}$

$= \dfrac{3 \pm \sqrt{3}}{3}$

71. $4x^2 + 4x = x^2 + 5$

$3x^2 + 4x - 5 = 0$

$x = \dfrac{-4 \pm \sqrt{4^2 - 4(3)(-5)}}{2(3)}$

$= \dfrac{-4 \pm \sqrt{76}}{6}$

$= \dfrac{-4 \pm 2\sqrt{19}}{6}$

$= \dfrac{-2 \pm \sqrt{19}}{3}$

73. $x^2 + 5x = -1 + 2x$

$x^2 + 3x + 1 = 0$

$x = \dfrac{-3 \pm \sqrt{3^2 - 4(1)(1)}}{2(1)}$

$= \dfrac{-3 \pm \sqrt{5}}{2}$

75. $\qquad 2y^2 = -3y - 1$

$2y^2 + 3y + 1 = 0$

$(2y + 1)(y + 1) = 0$

$2y + 1 = 0 \qquad y + 1 = 0$

$y = -\dfrac{1}{2} \qquad y = -1$

77. $3x^2 + 1 = 73 + x^2$

$2x^2 - 72 = 0$

$2x^2 = 72$

$x^2 = 36$

$x = \pm 6$

79. $\dfrac{(y - 2)^2}{20} + 3 + y = 0$

$(y - 2)^2 + 60 + 20y = 0$

$y^2 - 4y + 4 + 60 + 20y = 0$

$y^2 + 16y + 64 = 0$

$(y + 8)^2 = 0$

$y + 8 = 0$

$y = -8$

81. $\qquad 3x^2 + 1 = 6 - 8x$

$3x^2 + 8x - 5 = 0$

$x = \dfrac{-8 \pm \sqrt{8^2 - 4(3)(-5)}}{2(3)}$

$= \dfrac{-8 \pm \sqrt{124}}{6}$

$= \dfrac{-8 \pm 2\sqrt{31}}{6}$

$= \dfrac{-4 \pm \sqrt{31}}{3}$

83. $2y - 10 = 10y(y - 2)$

$2y - 10 = 10y^2 - 20y$

$0 = 10y^2 - 22y + 10$

$0 = 5y^2 - 11y + 5$

$y = \dfrac{-(-11) \pm \sqrt{(-11)^2 - 4(5)(5)}}{2(5)}$

$= \dfrac{11 \pm \sqrt{21}}{10}$

85. $\dfrac{y^2+5}{2y} = \dfrac{2y-1}{3}$

$3(y^2+5) = 2y(2y-1)$

$3y^2+15 = 4y^2-2y$

$0 = y^2-2y-15$

$0 = (y-5)(y+3)$

$y-5=0 \qquad y+3=0$

$\qquad y=5 \qquad \qquad y=-3$

87. $y = 2x^2$

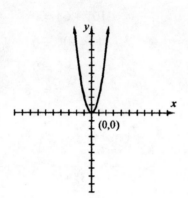

x	y
-2	8
-1	2
0	0
1	2
2	8

(0,0)

89. $y = x^2-3$

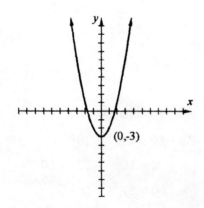

x	y
-3	6
-1	-2
0	-3
1	-2
3	6

(0,-3)

91. $y = x^2-3x-4$

Vertex: $x = \dfrac{-(-3)}{2(1)} = \dfrac{3}{2}$

$y = \left(\dfrac{3}{2}\right)^2 - 3\left(\dfrac{3}{2}\right) - 4 = -\dfrac{25}{4}$

$V\left(\dfrac{3}{2}, -\dfrac{25}{4}\right)$

x	y
-1	0
0	-4
1.5	-6.25
3	-4
4	0

(1.5,-6.25)

93. $y = -2x^2+12x-17$

Vertex: $x = \dfrac{-12}{2(-2)} = 3$

$y = -2(3)^2 + 12(3) - 17 = 1$

$V(3, 1)$

x	y
1	-7
2	-1
3	1
4	-1
5	-7

(3,1)

95. $L = $ length

$w = $ width

Length: $L + 2w = 11$ (1)

Area: $Lw = 15$ (2)

Substitute $\dfrac{15}{w}$ for L in (1)

$$\frac{15}{w} + 2w = 11$$

$$15 + 2w^2 = 11w$$

$$2w^2 - 11w + 15 = 0$$

$$(2w - 5)(w - 3) = 0$$

$2w - 5 = 0$ $w - 3 = 0$

$w = \dfrac{5}{2}$ $w = 3$

$L = \dfrac{15}{w} = 6$ $L = \dfrac{15}{w} = 5$

The dimensions are 5 feet by 3 feet or 6 feet by 2.5 feet.

97. $x = $ number of members last year

$x - 4 = $ number of members this year

Dues last year = dues this year

$720x = 726(x - 4)$

$720x = 726x - 2904$

$2904 = 6x$

$484 = x$

There were 484 members last year.

99. $w = $ width of rug

$w + 4 = $ length of rug

area of room – area of rug = area of border

$(w + 4)(w + 4 + 4) - w(w + 4) = 68$

$$w^2 + 12w + 32 - w^2 - 4w = 68$$

$$8w = 36$$

$$w = 4.5$$

$$w + 4 = 8.5$$

Area of rug $= (4.5)(8.5) = 38.25 \text{ ft}^2$

1. $5x^2 + 7x = 4$

$5x^2 + 7x - 4 = 0$

$a = 5, \ b = 7, \ c = -4$

$$x = \frac{-7 \pm \sqrt{7^2 - 4(5)(-4)}}{2(5)}$$

$$= \frac{-7 \pm \sqrt{129}}{10}$$

3. $2x^2 = 2x - 5$

$2x^2 - 2x + 5 = 0$

$a = 2, \ b = -2, \ c = 5$

$$x = \frac{-(-2) \pm \sqrt{(-2)^2 - 4(2)(5)}}{2(2)}$$

$$= \frac{2 \pm \sqrt{-36}}{4}$$

There is no real solution.

5. $12x^2 + 11x = 5$

$12x^2 + 11x - 5 = 0$

$(4x + 5)(3x - 1) = 0$

$4x + 5 = 0$ $3x - 1 = 0$

$4x = -5$ $3x = 1$

$x = -\dfrac{5}{4}$ $x = \dfrac{1}{3}$

7. $2x^2 - 11x + 3 = 5x + 3$

$2x^2 - 16x = 0$

$2x(x - 8) = 0$

$2x = 0$ $x - 8 = 0$

$x = 0$ $x = 8$

9. $2x(x - 6) = 6 - x$

$2x^2 - 12x = 6 - x$

$2x^2 - 11x - 6 = 0$

$(2x + 1)(x - 6) = 0$

$2x + 1 = 0$ $x - 6 = 0$

$x = -\dfrac{1}{2}$ $x = 6$

11. $y = 3x^2 - 6x$

Vertex: $x = -\dfrac{-6}{2(3)} = 1$

$y = 3(1)^2 - 6(1) = -3$

$V(1, -3)$

x	y
-1	9
0	0
1	-3
2	0
3	9

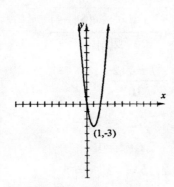

(1,-3)

13.

$x = $ length of one leg

$x + 3 = $ length of other leg

$x^2 + (x + 3)^2 = (15)^2$

$x^2 + x^2 + 6x + 9 = 225$

$2x^2 + 6x - 216 = 0$

$x^2 + 3x - 108 = 0$

$(x + 12)(x - 9) = 0$

$x + 12 = 0 \qquad x - 9 = 0$

$x = -12 \qquad x = 9$

Length can't be negative, so

$x = 9$ and $x + 3 = 12$.

The lengths are 9 m and 12 m.

Cumulative Test - Chapters 1 - 10

1. $3x\{2y - 3[x + 2(x + 2y)]\}$

$= 3x\{2y - 3[x + 2x + 4y]\}$

$= 3x[2y - 3(3x + 4y)]$

$= 3x[2y - 9x - 12y]$

$= 3x(-9x - 10y)$

$= -27x^2 - 30y$

3. $\dfrac{1}{2}(x - 2) = \dfrac{1}{3}(x + 10) - 2x$

$3(x - 2) = 2(x + 10) - 6(2x)$

$3x - 6 = 2x + 20 - 12x$

$13x = 26$

$x = 2$

5. $(3x + 1)(2x - 3)(x + 4)$

$= (6x^2 - 9x + 2x - 3)(x + 4)$

$= (6x^2 - 7x - 3)(x + 4)$

$= 6x^3 + 24x^2 - 7x^2 - 28x - 3x - 12$

$= 6x^3 + 17x^2 - 31x - 12$

7. $y = -\dfrac{3}{4}x + 2$

slope $= -\dfrac{3}{4}, \quad y - $ intercept $= 2$

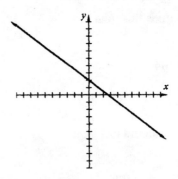

9. $3x + 2y = 5 \qquad (1)$

$7x + \ y = 19 \qquad (2)$

Multiply (2) by -2

$\qquad 3x + 2y = 5$

$\underline{-14x - 2y = -38}$

$-11x \qquad\ = -33$

$\qquad x = 3$

Substitute 3 for x in (1)

$3(3) + 2y = 5$

$\qquad 2y = -4$

$\qquad\ y = -2$

$(3, -2)$

210

11. $\sqrt{18x^5y^6z^3} = \sqrt{9 \cdot 2x^4xy^6z^2z}$
$= 3x^2y^3z\sqrt{2xz}$

13. $\dfrac{\sqrt{5}-3}{\sqrt{5}+2} = \dfrac{\sqrt{5}-3}{\sqrt{5}+2} \cdot \dfrac{\sqrt{5}-2}{\sqrt{5}-2}$
$= \dfrac{\sqrt{25}-2\sqrt{5}-3\sqrt{5}+6}{\left(\sqrt{5}\right)^2-2^2}$
$= \dfrac{11-5\sqrt{5}}{5-4}$
$= 11-5\sqrt{5}$

15. $H = 25b^2 - 6$
$H+6 = 25b^2$
$\pm\sqrt{H+6} = \sqrt{5^2b^2}$
$\pm\sqrt{H+6} = 5b$
$\dfrac{\pm\sqrt{H+6}}{5} = b$

17. $(2x+1)^2 = 20$
$2x+1 = \pm\sqrt{20}$
$2x = -1 \pm 2\sqrt{5}$
$x = \dfrac{-1\pm 2\sqrt{5}}{2}$

19. $3x^2 + 11x + 2 = 0$
$a = 3, \ b = 11, \ c = 2$
$x = \dfrac{-11 \pm \sqrt{11^2 - 4(3)(2)}}{2(3)}$
$= \dfrac{-11 \pm \sqrt{97}}{6}$

21. $x^2 = 98$
$x = \pm\sqrt{98}$
$x = \pm\sqrt{49 \cdot 2}$
$x = \pm 7\sqrt{2}$

23. $y = x^2 + 6x + 10$

Vertex: $x = \dfrac{-6}{2(1)} = -3$

$y = (-3)^2 + 6(-3) + 10 = 1$

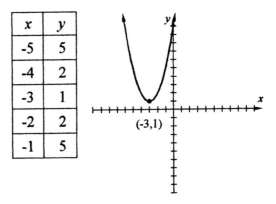

x	y
-5	5
-4	2
-3	1
-2	2
-1	5

(-3,1)

Practice Quiz: Sections 10.1 - 10.2

1. Solve by factoring. $x^2 - x - 6 = 0$

2. Solve by factoring. $5x^2 - 17x - 12 = 0$

3. Solve by taking the square root.
$x^2 - 18 = 0$

4. Solve by taking the square root.
$(x+3)^2 - 11 = 0$

5. Solve by completing the square.
$2x^2 - 3x - 4 = 0$

Practice Quiz: Sections 10.3 - 10.5

In Exercises 1 - 3 solve if possible
by using the quadratic formula.

1. $3x^2 - x - 4 = 0$

2. $x^2 + 2x - 11 = 0$

3. $3x^2 - 2x + 12 = 0$

4. Graph $y = x^2 + 2x - 3$.
Locate the vertex.

211

5. The diagonal of a rectangle is 25 m. The length of the rectangle is 3 more than 3 times the width. Find the dimensions of the rectangle.

Answers to Practice Quiz

Sections 10.1 - 10.2

1. $(-2, 3)$

2. $\left(-\dfrac{3}{5}, \dfrac{1}{4}\right)$

3. $x = \pm 3\sqrt{2}$

4. $x = -3 \pm \sqrt{11}$

5. $x = \dfrac{3 \pm \sqrt{41}}{4}$

Answers to Practice Quiz

Sections 10.3 - 10.5

1. $\left(-1, \dfrac{4}{3}\right)$

2. $x = -1 \pm 2\sqrt{3}$

3. No real solution.

4.

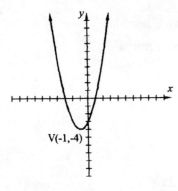

V(-1,-4)

5. 24 m by 7 m

1. $-2x + 3y\{7 - 2[x - (4x + y)]\}$

$= -2x + 3y[7 - 2(x - 4x - y)]$

$= -2x + 3y[7 - 2(-3x - y)]$

$= -2x + 3y(7 + 6x + 2y)$

$= -2x + 21y + 18xy + 6y^2$

3. $(-3x^2 y)(-6x^3 y^4) = 18x^5 y^5$

5. $\dfrac{1}{2}(x + 4) - \dfrac{2}{3}(x - 7) = 4x$

$3(x + 4) - 4(x - 7) = 6(4x)$

$3x + 12 - 4x + 28 = 24x$

$40 = 25x$

$\dfrac{8}{5} = x$

7. $5x + 3 - (4x - 2) \le 6x - 8$

$5x + 3 - 4x + 2 \le 6x - 8$

$-5x \le -13$

$x \ge \dfrac{13}{5}$

←———+———●———→
       0   13/5

9. $4x^2 - 18x - 10 = 2(2x^2 - 9x - 5)$

$= 2(2x + 1)(x - 5)$

11. $\dfrac{2}{x-3} - \dfrac{3}{x^2 - x - 6} + \dfrac{4}{x+2}$

$= \dfrac{2}{x-3} \cdot \dfrac{x+2}{x+2} - \dfrac{3}{(x-3)(x+2)} + \dfrac{4}{x+2} \cdot \dfrac{x-3}{x-3}$

$= \dfrac{2x + 4 - 3 + 4x - 12}{(x+2)(x-3)}$

$= \dfrac{6x - 11}{(x+2)(x-3)}$

212

13. $\dfrac{2}{x+2} = \dfrac{4}{x-2} + \dfrac{3x}{x^2-4}$

$(x-2)(2) = 4(x+2) + 3x$

$2x - 4 = 4x + 8 + 3x$

$-12 = 5x$

$-\dfrac{12}{5} = x$

15. $m = -\dfrac{3}{4}, \ (-2, 5)$

$y = mx + b$

$5 = -\dfrac{3}{4}(-2) + b \Rightarrow b = \dfrac{7}{2}$

$y = -\dfrac{3}{4}x + \dfrac{7}{2}$ or $3x + 4y = 14$

17. $2x + 3y = 8 \qquad (1)$

$3x - 5y = 31 \qquad (2)$

Multiply (1) by 3 and (2) by -2

$6x + 9y = 24$

$-6x + 10y = -62$

$\overline{19y = -38}$

$y = -2$

Substitute -2 for y in (1)

$2x + 3(-2) = 8$

$2x = 14$

$x = 7$

$(7, -2)$

19. $2x\sqrt{50} + \sqrt{98x^2} - 3x\sqrt{18}$

$= 2x\sqrt{25 \cdot 2} + \sqrt{49 \cdot 2x^2} - 3x\sqrt{9 \cdot 2}$

$= 10x\sqrt{2} + 7x\sqrt{2} - 9x\sqrt{2}$

$= 8x\sqrt{2}$

21. $\dfrac{\sqrt{3} + \sqrt{7}}{\sqrt{5} - \sqrt{7}} = \dfrac{\sqrt{3} + \sqrt{7}}{\sqrt{5} - \sqrt{7}} \cdot \dfrac{\sqrt{5} + \sqrt{7}}{\sqrt{5} + \sqrt{7}}$

$= \dfrac{\sqrt{15} + \sqrt{21} + \sqrt{35} + \sqrt{49}}{\left(\sqrt{5}\right)^2 - \left(\sqrt{7}\right)^2}$

$= -\dfrac{7 + \sqrt{15} + \sqrt{21} + \sqrt{35}}{2}$

23. $2y^2 = 6y - 1$

$2y^2 - 6y + 1 = 0$

$y = \dfrac{-(-6) \pm \sqrt{(-6)^2 - 4(2)(1)}}{2(2)}$

$= \dfrac{6 \pm \sqrt{28}}{4}$

$= \dfrac{6 \pm 2\sqrt{7}}{4}$

$= \dfrac{3 \pm \sqrt{7}}{2}$

25. $c^2 = a^2 + b^2$

$8^2 = a^2 + 5^2$

$64 = a^2 + 25$

$39 = a^2$

$\pm\sqrt{39} = a$

The length can't be negative.

$a = \sqrt{39}$

27. $w = $ width

$2w - 2 = $ length

$p = 2L + 2w$

$38 = 2(2w - 2) + 2w$

$38 = 4w - 4 + 2w$

$42 = 6w$

$7 = w, \ 2w - 2 = 12$

The dimensions are 12 meters by 7 meters.

29. x = number of general admission tickets

y = number of reserved seat tickets

$x + y = 360$ (1)

$3x + 5y = 1480$ (2)

Multiply (1) by -3 and add it to (2)

$2y = 400$

$y = 200$

Substitute 200 for y in (1)

$x + 200 = 360$

$x = 160$

200 reserved seat tickets

160 general admission tickets